Unpacking Construction
Site Safety

Unpacking Construction Site Safety

Dr Fred Sherratt
MCIOB C.BuildE MCABE FHEA

WILEY Blackwell

Library of Congress Cataloging-in-Publication data applied for

ISBN: 9781118817285

A catalogue record for this book is available from the British Library.

Wiley also publishes its books in a variety of electronic formats. Some content that
appears in print may not be available in electronic books.

Set in 10.5/12.5pt Avenir by SPi Global, Pondicherry, India

1   2016

For Pru

Contents

Contents

ix

Preface

I have worked in the construction industry for over 13 years. I began as a site secretary and worked my way up through the ranks via the planning function to site management. It is an industry full of interesting, entertaining and wonderful people who all make something happen. It is an industry that creates things, that makes places and spaces for people and changes the world we live in. Whilst sometimes fraught with conflict, aggravation and traumas, it is also an industry full of life and laughter and usually someone singing very loudly, a little bit off key. It is an industry that I love.

But it also has a big problem. I have seen the consequences of accidents that have stopped men working for weeks and months. I have had to collect the witness statements and take the photographs of the locations when accidents have occurred. I have had to gather the evidence that they had been inducted and read their method statements for the task they were performing at the time. I have donated to collections to try to keep a family going as no income will be forthcoming for the next few months whilst an injury heals and bills still have to be paid.

My position within this environment enabled me to approach health and safety in a different way from many of my peers. As a woman on site I was different, although I never felt that I did not fit in; I found that construction accepts you if you can do the job you are there to do, no matter what gender, race or age you are. I am able to swear with the best of them, shout when shouting is needed, and coax and persuade when required. And because of this perhaps, I was able to argue from the point of view of the wife or daughter, I was able to show concern where my colleagues resorted to anger, I was able to suggest that the consequences might outweigh the benefits, I was able to say that I was stopping work because I cared. And when this approach was articulated it did make a difference, and people did listen.

However, this did not manage to stop people acting unsafely. Every day I saw that people did not follow the rules, despite inductions and training they still acted unsafely, they still took risks and they still did not always behave with care and concern for everyone else on the site. I sat in training rooms with them on safety culture training programmes, which took a different approach to safety, and I heard the comments afterwards, not to mention the comments before, that they were to lose half a day's pay for this 'shite'.

And that is what initiated my research, once I'd finished my degree in construction management I carried on. I wanted to ask the question why, despite best efforts all round, and the agreement that things could still improve in terms of safety (although some of the training left a lot to be desired), did accidents and incidents still occur? Why were we still having collections? Why did you still hear stories and tales of accidents on other sites, of the deaths of people that we had been working alongside only a few months ago? Why in the twenty-first century had this not yet been sorted out?

Alongside my working life spent living the construction dream, I am also a geek. I like to study and explore and think about things, to learn about new ideas and approaches. I could see that most of the ways we tried to measure safety weren't working; safety climate questionnaires were completed with what the management wanted to hear, not a reflection of reality. This was also the case in academia, where research often measured people as if they were constant, that they could be predicted, that they behaved according to rules and logical thought. Reality tended to argue quite strongly with this. I wanted to know why this didn't work, or rather, didn't seem to me to work? What alternatives for exploration existed? Could they help? Could they provide a different perspective on people and help us understand how to make it safer on sites?

Consequently, I started at Plato and carried on. I discovered cognitive theories and became very excited, I applied this thinking to risk taking on sites, and it kind of worked, but didn't really tell me anything that had not been found out before. And as I kept investigating, I found that maybe this approach couldn't answer all the questions in terms of my experience. It couldn't predict or explain everything that was common in terms of the uncommon found on sites, and when it tried it tied itself in paradoxical knots. I kept going, and found social constructionism which through its approach didn't even try to explain. It enabled acceptance and understanding rather than any 'scientific' explanations. It unquestionably embraced variation, irrationality, and crazy stupid people doing crazy stupid things, without trying to explain them. It let you explore and understand, without the need for assumptions or generalisations. As far as I could establish it hadn't ever been used on construction sites; this approach hadn't been tried before. Maybe it could throw out some new ideas, some new suggestions that could help?

I could see that it might not provide the answers that people who write training programmes might want to hear. It didn't produce firm explanations which could be located in the crosshairs and eliminated from sites. Rather it offered insight, illumination and understanding. More thinking would be required once this was achieved, but I wanted to see where this path led. So off I went.

Acknowledgements

I would like to thank Dr Peter Farrell and Dr Rod Noble for their support and guidance throughout this project – and to Roger Seeds for providing the very first encouragement. Also, thanks must go to Paddy O'Rourke and Julia Stevens for the opportunity to make it happen. And to Barry Rawlinson for his time, efforts and honesty – as always – and yes Egg, this really, really is it!

Chapter One
Introduction

This book aims to explore and unpack construction site *safety*. From the very start it must be made clear that this does not include its long-time associate *health*, or the more recent addition of *well-being*. The reasons for this will quickly become apparent, but are broadly due to differences in the way they emerge on sites, how they are managed in practice, and in part their very essence. As will be examined later, there are fundamental differences between them that should arguably be better acknowledged and considered within construction management, yet for this text they have been set to one side in order to ensure full attention can be paid to the specific concept of construction site *safety*. However, health in particular does still appear in general contextual discussions, placed alongside safety as part of a seemingly unbreakable, although at times impractical and often not very helpful, amalgam.

This book takes a different approach to safety on construction sites.

Rather than discussing the implementation of various regulations or seeking to evaluate the effectiveness of safety management systems against templates of 'best practice', it considers *how* people think about safety, what it means to them and how they go on to collectively use those ideas in their everyday work. This could also be deemed an evaluation of construction site 'safety culture', a notoriously problematic term and one that is discussed in more

Unpacking Construction Site Safety, First Edition. Dr Fred Sherratt.
© 2016 John & Wiley Sons, Ltd. Published 2016 by John & Wiley Sons, Ltd.

detail in later chapters.Although to some extent, that is precisely what this book is.

This book takes the approach of asking some very fundamental questions.

What is safety on site?
Do we agree on our definition?
How do we talk about it?
How is safety associated with practice?
Does it 'work'?

Although the last question has already been partly answered for us by the fact that we keep appearing in the list of the UK's most dangerous industries, it, and these other questions, will be explored as construction site safety is unpacked within this book.

The term 'unpacking' may seem a little odd. It comes from the way this book has been researched and prepared. It means to pull apart, to challenge, to question and to consider from as wide a variety of perspectives as possible, both academic and practice-based. It therefore lets us take safety apart within the specific construction site context to see what we can find – an ideal approach to help us answer the questions above, allowing us to explore and address them from outside the traditional frameworks of legislation, management systems and best practice. Instead, we can see how these approaches actually work in practice, how they are received by those who have to use them on a daily basis, and how they ultimately contribute to what safety actually *is* on sites. The way this process has been carried out is discussed in much more detail in Chapter 3.

The context for this book is large UK construction sites (over £15 million in value) operated by large main contractors (found within the top 30 contractors in terms of annual work won by value in the UK), rather than those operated by small-to-medium sized enterprises (SMEs) or micro operations and sole traders. However, smaller industry organisations inevitably participate in work on large sites as they operate as subcontractors within industry supply chains. Research has shown that subcontractors take their ideas of safety with them when they move from project to project (Aboagye-Nimo *et al.* 2012), and therefore SMEs and even micro-SMEs play a considerable part in helping to create and perpetuate what safety is on large construction sites.

Within the contemporary UK construction industry, main contractors can be seen to be actively trying to improve their safety management through the use of structured Safety Management Systems, a focus on accident targets and various safety management programmes. In this environment, such well-implemented safety management should ideally mean zero accidents, but it doesn't. Sadly there are still incidents on large projects; the death of a worker in March 2014 on Crossrail in London occurred despite a certified safety management system and Target Zero safety programme being in place (Crossrail 2015). These environments are where 'traditional' safety management has been suggested to have plateaued in terms of what it can achieve, and so where new thinking is needed for future improvements.

Reading this book will hopefully support the development of a deeper understanding of safety on sites, which goes beyond practical frameworks of legislation and management systems, and starts to consider the answers to the questions asked earlier in detail. With a better knowledge of how safety actually 'works' within the site context, the development and implementation of management systems, interventions and initiatives can be subsequently enhanced and tailored to improve 'fit' within this environment. There is also the potential to improve existing safety management practices, by enabling a better understanding of *why* people might sometimes act as they do when they carry out safety violations, enabling the best course of action to be determined, both with the individual (to engage and educate or to discipline and punish) but also within the wider work context (to change the work method or revise payment practices, for example).

This book is intended for practitioners, academics and students of construction management. It hopes to cross the divide between practice and academia, both of which need each other to gain a complete picture of any aspect of construction management. Where some elements of this book will necessarily explore how we think about things and what this means for our social interactions from academic perspectives, there is also the need to illustrate and explain these academic considerations in relevant and representational contexts of practice.

Although the author is now works as an academic, she has over 10 years' experience of working on large construction sites in the UK, including several years as a construction section manager. During this time she was directly involved in safety, and has therefore worked through the challenges of its implementation, as well as unfortunately been witness to the repercussions when it has sadly failed.

This book seeks to draw on both academia and practice, and it is hoped that from either perspective, the other viewpoint proves illuminating and that both can be brought together here to give a different, informative and most importantly *useful* understanding of safety on construction sites.

References

Aboagye-Nimo, E., Raiden, A., Tietze, S. and King, A. (2012) The use of experience and situated knowledge in ensuring safety among workers of small construction firms. In S.D. Smith (ed.), *Proceedings 28th Annual ARCOM Conference*, pp. 413–22. Association of Researchers in Construction Management, Edinburgh.

Crossrail (2015) *Health and Safety* [Online]. Available: http://www.crossrail.co.uk/sustainability/health-and-safety/ [30 March 2015].

Chapter Two
Construction Site Contexts

Our job, they say, is to get stuck in and get the job done, not to fill in forms. In time this macho approach becomes the local custom and practice.

Kletz 2012: 765

Although Kletz was not specifically talking about the UK construction industry when he made this statement, he might as well have been. Getting stuck in and getting the job done can be seen as one of our industry's most positive characteristics – nothing can't be done! – but it has also arguably contributed to one of the worst safety records in UK occupational safety.

The Health and Safety Executive (2014) report that the UK construction industry only employs approximately 5% of the UK workforce, but disproportionately accounts for 31% of fatal injuries, 10% of reported major/specified injuries and 6% of over-7-day injuries to employees. In the period 2013/14 there were 42 fatal injuries to workers in the construction industry and 592 000 working days were lost due to workplace injury, a total of 1.1 days lost per worker. All these statistics make for unpleasant reading, and also make construction one of the most dangerous industries to work in within the UK.

Often accidents happen because of changes to planned work, something pretty much inevitable in the construction industry.

Unpacking Construction Site Safety, First Edition. Dr Fred Sherratt.
© 2016 John & Wiley Sons, Ltd. Published 2016 by John & Wiley Sons, Ltd.

We build our own work environments around us and so bring change to our workplaces on a daily basis – if we didn't we wouldn't be doing our job – but this is something no other industry really has to contend with. For example, the management of access routes around a construction site can be a very complex and time-consuming task – if the stairs used to access the third floor yesterday are being screeded today and so everyone needs to go round outside to the door at the bottom of the next staircore, but not round the east as the curtain walling is going up and there's no access through, but that will change next week when they drop onto the west … and so on. And change is not limited to the physical workspace; change to programme, to sequence, to design, to work practices and methods can also occur on a fairly regular basis as labour and plant become available or unavailable, or our clients simply change their minds. As a result the construction industry is highly accepting of change, and sees it as an inherent part of work.

But changes in work environments can also make significant changes to the hazards and risks of a task, and in such cases change means safety should be reconsidered and re-planned and reprogrammed. But these safety aspects can go unnoticed or even ignored, because getting the job done is our top priority. And that is when accidents can occur.

This flexible and fluid work context is also influenced by other aspects of the industry: the motivations behind getting the job done, the people who carry out the work, the way work is allocated and paid for and even the working conditions. Understanding of this wider context provides the groundwork for understandings of how safety itself works on sites, and the contextual influences that have shaped it within this construction site environment.

Winning Work

Winning work in construction can be a complex process, not least because of significant variations in work availability. Demand for construction work is directly derived from the needs of other industries or the public sector (Morton and Ross 2008). Given the nature of the product and the need for capital expenditure or investment for its production, this demand is closely linked to the overall health of the UK economy, and the industry goes through boom and bust periods as the economy fluctuates between growth and recession (Dainty *et al.* 2007). As the economic downturn of 2010 has demonstrated, the construction industry can be hit hard in terms of workload reduction and job losses during recession, only to be short of materials, labour and skills to carry out projects once the workload picks up. Lengthy project timescales can also mean that work priced and won in a time of recession can also hit problems when the start on site coincides with a recovering market – labour and material prices rise and so place a squeeze on profit margins that may well have been tight to begin with.

Construction work is traditionally won through competitive tendering processes with the award of work usually going to the lowest bidder. This makes organisational workloads highly uncertain, and means companies are under pressure to keep their bids low to increase their chance of winning. This can inevitably lead to a focus on cost rather than other project considerations, such as quality, sustainability and of course safety (Lingard and Rowlinson 2005). This can even be the case in more 'balanced' tenders such as those found in partnering agreements or frameworks; although quality or sustainability or safety are more likely to be acknowledged here, price often remains the factor with most influence when work is awarded.

In addition to cost, time is also critical – not least to ensure construction companies do not overrun the agreed contract duration and incur additional costs themselves (Loosemore et al. 2003). Clients will also consider project duration when awarding their work, and so companies also frequently bid for work with promises of delivery within very short timescales.

As a result, productivity is king and the two driving forces of time and money filter down from clients, through the project and site management teams, to the operatives carrying out work on site. Speed is of the essence; there is a constant pressure to meet daily or weekly targets on sites, be it real or perceived, which forms an ingrained aspect of construction site life (Health and Safety Executive 2009a). The tight profit margins necessitated by competitive tendering can lead to complex value engineering and a reliance on inexpensive working methods. In essence, work must be carried out as quickly and cheaply as possible. Although change has been forthcoming with the advent of partnering and other collaborative working practices, it is the two factors of time and money that still form the bottom line of the vast majority of construction projects. It is therefore unsurprising that these two elements have arguably become ingrained as dominant 'truths' within the site community.

These pressures have also contributed to the perception of construction sites as places of antagonism, with conflict described as 'institutionalised' within the industry (Loosemore *et al.* 2003). Many reports have berated the adversarial and antagonistic aspects of the industry which have led to an aggressive, conflict-ridden environment (Watts 2007). Several reasons have been presented for this. The project-based nature of the work has been blamed, as organisations come together on a project-by-project basis, with differing and occasionally competing objectives and demands (Fryer *et al.* 2004). The payment processes of the industry have also been cited as problematic – that our payment practices have even required a law (The Construction Act) to ensure fairness is more than a little embarrassing – and that the competitive tendering process simply leads to a 'claims culture' seeking variations and additional payments from the client, once the work has been won on a low tender price (Rooke *et al.* 2004). At the site level, the use of differing trades within the supply chain also results in competing objectives; each trade wants to complete its work efficiently, but a reliance on the success of the previous trade, competition with others to complete their work first in an area, and disagreements in the proposed planning of the work can all result in conflict on sites.

Subs of Subs of Subs

As a result of the inherent uncertainties in work winning and subsequent variations in workload, construction companies require a high degree of flexibility to be able to cope with fluctuations. Consequently, subcontracting of work is prolific and has become the dominant organisational structure for large construction projects (Dainty *et al.* 2007). Main contractors win the work through the tendering process, and then assign packages of work dependent on trade or skill to many different subcontractors in their supply chain, again through a competitive tendering process. These subcontractors can also subcontract work, resulting in elongated supply chains and highly fragmented delivery systems, often with the pressures and risks of time and cost being cascaded down to the levels below. Main contractors are unlikely to have any direct authority over the subcontractors' operatives (Fryer *et al.* 2004), which often results in hierarchical systems of management; from the main contractors' management to their supervisors to the subcontractors' supervisors to the subcontractors' operatives, with levels of responsibility and accountability all clearly defined (Watts 2007). Yet this potentially beneficial flexibility has also been criticised as it creates conflicting interests on site by subdividing the project (Ankrah *et al.* 2007), as well as increased health and safety concerns due to poor house-keeping and a lack of effective safety training, which can increase accidents on sites (Lingard and Rowlinson 2005).

The need for flexibility also translates to the workforce, with a significant amount of construction operatives being self-employed (Dainty *et al.* 2007), in the real or bogus sense of the term. However, such casual labour practices also have negative consequences for the industry and the pressures of time and money are of course very immediate concerns of self-employed operatives who have to work to earn, as their contractual arrangements disoblige employ-ers from statutory responsibilities such as holiday and sick pay. This arrangement also releases employing companies from any responsibilities for training such operatives, including training and qualifications for safety (Morton and Ross 2008).

For the self-employed, and even those working directly for con-tractors, the common practice of paying on 'price' or 'measure' – the amount of work carried out in the day – adds yet another pressure. Payment on price is frequently used as an incentive pay-ment scheme to increase productivity, facilitated by the ease with which outputs can be measured and rewarded (Harris *et al.* 2006). However, this practice inevitably encourages operatives to work as fast as possible to make the most money in a day or shift, or even worse creates a situation in which operatives *have* to work as fast as possible to make any profit on a job that has had it all squeezed out of it all the way along the supply chain. As speed often means cutting corners and taking risks, safety is often sacrificed (Spanswick 2007). The ever looming deadline for completion of projects means there is constant pressure to meet daily and weekly targets. This pressure is often most keenly felt by site foremen, supervisors and site managers, who often turn a blind eye to unsafe practices with fingers crossed, to achieve the necessary production (Health and Safety Executive 2003).

The Workforce

The very nature of construction work has inevitably created a project-based industry, where temporary project teams are formed on the construction sites. The workforce comes together for the duration, only to disband at project completion to start again elsewhere (Sang *et al.* 2007). Many large construction companies are structured so their projects, or sites, are self-contained, autonomous entities, able to manage their own costs and profits as the project managers or leaders see fit, allowing individual sites to develop their own 'site culture'. This structure has inevitably led to the creation of a transient workforce (Health and Safety Executive 2009a), with high levels of casual recruitment and short-term work contracts (Haro and Kleiner 2008) as the operatives move from project to project. It has been argued that this itinerant workforce has repercussions for the work itself, that the very nature of the employment promotes a casual attitude to the work, and a workforce that does not accept conventions on punctuality, attendance and safety that apply to more regular work (Seymour and Fellows 2002).

In terms of skills, the construction industry has historically had a very low competence threshold for site-based operatives, in part encouraged by the short-term and itinerant nature of the work which can make long-term training a problem (Health and Safety Executive 2009a). The industry instead often looks to knowledge and experience as benchmarks for competence over formal qualifications (Rooke and Seymour 2002), with qualifications considered an irrelevant measure of people's actual skills and ability. Employment within the construction workforce is usually based on word-of-mouth referrals and informal recruitment networks, with associated operatives often travelling from project to project together, supporting each other in finding future work. However, this recruitment process excludes as many people as it includes, and has ultimately resulted in the perpetuation of the white male domination of the workforce (Ness 2009).

Less than 1% of the construction industry operative and site-based workforce are women (UCATT 2015), and it has been estimated that less than 4% within the workforce as a whole are from a black or ethnic minority background (Chaudhry 2014). Whilst the industry has been a traditional employer of foreign and migrant workers on sites, they have been estimated to only form around 12% of the site-based workforce (McMeeken 2015). The vast majority of the site-based workforce is white and male. The lack of women within the workforce has led to what is frequently described as a 'macho culture' on sites, and on UK construction sites at least, this last bastion of the traditional male working class is characterised by the use of sexual language and humour, macho behaviour and almost constant swearing (Jordan *et al.* 2004). There are two theories as to why sites have developed in this way. One argues that the 'spirited conversation kept the wheels of productivity turning' (Gregory 2006) and such shared social behaviours allow for strong bonds to be formed quickly as workers are shifted round the site or from project to project, creating a sense of support and belonging within the workforce (Bird 2003). The second theory argues that the boisterous masculine culture of the male workplace can also be seen as a display of the workers' culture of resistance against capitalism which threatens to emasculate them (Cockburn 1983; Gregory 2006). The need to be tough and physically superior to their managers is one way the workers can compensate for the masculine 'mutation' of subordination to other men (Cockburn 1991).

This workforce composition and somewhat inevitably labelled 'macho culture' has also been suggested to have led to an acceptance of risk-taking behaviours on sites. Risk -taking can form part of the individual's construction of the masculine self (Lupton 2003), and in an industry whose sites are 99% male (UCATT 2015) this will inevitably have some influence. Indeed, the World Health Organisation (2002) concluded that 'masculinity may be hazardous to health', when it established a clear correlation between masculinity and risk taking. It has been suggested that men define themselves largely through their work (Cockburn 1991), which indicates that risk-taking behaviours are more likely to be carried out in this environment, and a construction site certainly provides the opportunity for this. Risk taking also helps to confirm an individual's independent choice and the control of their own lives, and so risk taking will also appeal to construction operatives who have been found to enjoy the high degree of autonomy that is part of the standard working arrangement (Applebaum 1981).

And construction operatives do take risks, as evidenced in industry-specific accident models. For example, two of the three root causes of accidents under the construction industry-specific Accident Root Cause Tracing Model (ARCTM) developed by Abdelhamid and Everett (2000) are assigned to deliberate risk taking on the part of operatives. Whether they decided to proceed with work activities once an existing unsafe condition had been identified, or decided to act unsafely regardless of the initial conditions of the work environment, either approach requires the identification of the risk – and its subsequent acceptance on behalf of the workforce.

This workforce composition has also been suggested to have contributed to the existence and perpetuation of other prominent characteristics of life on UK construction sites. Working hours on UK construction sites are often described as excessively long – operatives work some of the longest hours in Europe (Clarke *et al.* 2004). The male workforce, excused family responsibilities, bears the brunt of the pressures created by the work-winning process, with 'face-time' – a measure of commitment and productivity (Watts 2007) in what has also been described as a 'martyr culture' (Knutt 2009). The argument that such hours are necessary due to the 'nature' of the work is easily challenged; research in Holland, where shorter hours are the norm, found its productivity to be far higher than the UK (Clarke *et al.* 2004). The UK is, in fact, the only member of the EU to retain the right for an employee agreed exemption from the 1993 European Working Time Directive that set a weekly limit for 48 hours paid work to protect employees. This would suggest that even at a UK government level, long hours are seen as desirable commitment, despite the potential for safety problems to increase within an increasingly exhausted workforce.

There is also a continued reliance on manual labour found within the industry, stereotyped by the big, muscle-bound construction worker. Arguably necessary many years ago, the perpetuation of manual labour within a macho culture, where any indication of not being tough enough for the job is seen as a sign of weakness, has led to the construction industry having some of the highest levels of illness amongst its workers. The industry has around twice as many workers diagnosed with musculoskeletal disorders (MSDs) within the UK than in other industries (Health and Safety Executive 2014), the majority of which are back injuries from poor manual handling.

It can be suggested that work develops to suit its workforce, which in turn perpetuates these requirements in those it seeks to employ in the future. So it is perhaps unsurprising that the dominance of white men on construction sites has continued, despite many attempts to improve equality and diversity within the workforce. As a result, aspects such as risk tolerance and accepted ways of working, such as long hours and heavy manual labour, all of which can be readily associated with a 'macho culture', will have influence on how we manage our construction sites, as well as workers' expectations of their environments.

Working Conditions

Our workforce is also asked to carry out its tasks within some unique working conditions, which are highly influenced by the weather (Watts 2007). Depending on the stage of the work, it is the weather which determines if the operatives will be wet and cold all day, trudging through mud to get to the workface, or sweltering in the heat and dust, with the risk of sunburn. This gives support to the common perception that construction sites are dirty places with generally poor working conditions (Watts 2007). Indeed, Egan (1998) in his review of the industry described the site environment as 'challenging' in terms of the conditions found there. Due to the intensive use of heavy equipment and portable power tools, site operatives are frequently exposed to high levels of noise, which are often above allowable legal limits. The atmosphere on sites can also be unpleasant, with a wide range of construction processes, such as chasing, scabbling, drilling, crushing, cutting or breaking, raising silica dust and other particulates (Health and Safety Executive 2006a) making the very air of the site both dusty and hazardous.

The workplace itself can also be of concern, over and above the constant change that is an inevitable consequence of the work itself. That we are essentially an assembly business means materials, tools and people have to come together, often in relatively small spaces, in various ways to produce the finished product – and this can be messy. The Health and Safety Executive's 'Good Order Initiative' (2006b) was established to pass on the message that 'it is not acceptable for corridors and stairwells to be obstructed with materials, footpaths to be uneven, cables to be strewn across walkways or for steps into site cabins to be poorly constructed'. There are a variety of aspects inherent in the work that can contribute to poor working conditions, although it has been recognised that significant improvements have been made over recent years where practical or indeed possible (Health and Safety Executive 2003).

Never work a day in your life …

However, whilst the above paints a predominately negative picture of the construction industry, it is certainly not all bad. Very high levels of job satisfaction can be found on construction sites (see for example the research carried out by Jordan *et al.* 2004; Coffey and Fowler 2010; Polesie 2010). Job satisfaction can be drawn from many sources: pride in the participation and creation of something tangible (Watts 2007; Rawlinson and Farrell 2010; the empowerment and autonomy of the workforce (Applebaum 1981; Polesie 2010); team working (Jordan *et al.* 2004); the practical use of craft skills (Eisenberg 1998) and the satisfaction inherent in overcoming the many problems that can arise (Court and Moralee 1995; Chan and Kaka 2007). When I worked on construction sites I loved it. I loved it because I got to be part of a team, I got to work alongside some really interesting people as we planned and put things together, as we overcame challenges, and eventually delivered something that actually made a physical difference to the city I lived in.

So whilst the work environment of construction sites can be seen as hard and challenging, it is also a place of enjoyment, satisfaction, pride, laughter, and usually someone somewhere singing loudly and, unfortunately, quite badly off key.

Construction Site Life

Cipolla *et al.* (2006) argue that the very nature of the construction industry is the cause of its poor safety record. Some of the industry's key characteristics have been cited again and again as the root cause of many accidents and incidents: competitive tendering for work winning (Health and Safety Executive 2001; Sang *et al.* 2007; Morton and Ross 2008); the use of subcontracting and long supply chains (Donaghy 2009; Manu *et al.* 2010); the transient and fragmented workforce (Health and Safety Executive 2001; Biggs *et al.* 2005; Donaghy 2009); bonus and payment schemes that encourage speed and risk-taking behaviours (Sawacha *et al.* 1999; Gadd and Collins 2002; Fellows *et al.* 2002) and the constant demand for progress (Lingard and Rowlinson 2005; Health and Safety Executive 2009b). Indeed one-quarter of experts consulted for the Donaghy Report (2009) to the UK government – *One Death is Too Many* – felt that the way the industry is set up and work is procured has created an ethos that actively encourages safety accidents and incidents.

To fully understand the construction site context, an appreciation of what it can be like to work within this challenging environment will also be helpful. Consider the following …

7am: Morning Huddle

You have got to hit the milestone for the end of the week as your contracts manager is coming next Monday and will want to see results or know why but the morning meeting isn't going well as two of the subcontractors needed for the week's work already know each other from another project where one gang lost out in the money and so they both spend the meeting being rude to each other and refusing to agree on a plan for the week's work but you can't resolve it in the end as you get called out on site because a wagon has arrived with the materials you desperately needed yesterday which didn't turn up but this is now blocking the access road and a scissor lift is trying to get past but it breaks down halfway round and amongst all the abuse nothing is moving so you have to get down there to shout for the driver of the yellow van which is blocking the other side of the road but he's on the roof and is on his way but the wagon driver isn't too happy as he has five more drops today and makes you very aware of it and eventually the van is shifted to allow the wagon on but once you can get the forks to remove the first pallet you find the materials are the wrong colour for the job as the architect changed the design last week and for some reason this hasn't been passed on to the manufacturer whom when you ring him flatly states that any change will cost him money and he can't absorb that as he's still waiting payment from your company for last month's valuation but as soon as you escape from the phone two of your concrete lads are in your office and furious that their bonus is wrong from that pour last week that went on till eleven as the supervisor had promised them job and knock for a full second shift so this isn't right but the wages only run once a week so they'll have to make do till then and anyway you need to get back out on site to refix the access signs as today's concrete work will mean having to reroute the main entrance walkway but you can't get past the side door as two operatives are complaining about the joiners cutting MDF in the corridor and you know they have been allocated a room but they aren't carrying the timber down there as it takes too long and they're on price for this work so you have to issue some disciplinary notices to make sure they follow the rules which they do for now and then grab some masks for the lads working in the room next door as they are still worried about the dust but then you have to go and help your gen op carry the barriers across to reroute the pathway or he won't have time to refix them before the end of his shift as he's been on since half six to open up which makes your back start to twinge again but now a gang has turned up to connect up the rainwaters but the roofer hasn't finished yet so they give you an earful as they've driven down from Scotland and just found some lodge but they'll be back sometime next week or maybe the week after depending what's on but you're called back inside the building as no one can figure out how to access the high level sensor the M&E engineers want installing at the top of the atrium and when did this appear on the drawing as you only struck the scaffold last week and it wasn't there then but a phone call argues that it was on the drawing but only the electrical drawing not the BMS drawing which is what the sensor lads use so they'll need to source a scissor lift but the foreman doesn't think this should be up to them and you should have made sure everything was finished before the scaffold strike as they were able to use that for the other work so you stand in a four storey atrium and wonder if there is a spider that can get through the double doors and get that high and if there is have the lads got a ticket but no because the lad with a ticket is on holiday and not back for two weeks and by then the epoxy flooring should be down and it might get damaged with a spider and another phone call finds out that the sensor can go lower but the architect wanted it higher so as to not interfere with the design and so another phone call finds that the architect is in a meeting but will ring you back but this sensor needs to go in to finish the loop on the circuit or they can't test it and you can hear the saw going in the corridor again and the concrete lads' foreman is hovering in the doorway to try to get that wage thing sorted...*

*continues incessantly until 7pm when you eventually leave having said goodnight to your kids over the phone.

Again.

Summary

Some have argued that the construction industry operates within a unique work environment and construction managers face unique challenges. Others claim not, that it can be adapted and modified to something only a little more complicated than the production lines of manufacturing. However, it can also be argued that whatever the reasons behind them, these construction site contexts put significant pressures on the workforce even before they set foot in a workplace of leading edges, heights, excavations, plant and machinery, tools, chemicals and many more tangible hazards and direct risks to their safety.

Yet there is also one further consideration to be made to provide a fully holistic picture of safety on sites; the wider social context of safety will also have influence. Alongside the pressures of construction work and the immediate hazards within the working environment, what people read in the papers on their way into work, the jokes they tell in the pub, and the stream of posts popping up on their smartphone feeds will also contribute to and shape how people think about safety, and the wider ideas of safety and society should therefore also be acknowledged.

References

Abdelhamid, T. and Everett, J. (2000) Identifying root causes of construction accidents. *Journal of Construction Engineering and Management*, **126**(1), 52–60.

Ankrah, N., Proverbs, D. and Debrah, Y. (2007) A cultural profile of construction project organisations in the UK. In D. Boyd (ed.), *Proceedings of the 23rd Annual ARCOM Conference*, pp. 169–79. Association of Researchers in Construction Management, Belfast.

Applebaum, H.A. (1981) *Royal Blue: Culture of Construction Worker*. Holt, Rinehart and Winston, Boston.

Biggs, H., Dingsdag, D., Sheahan, V.L., Cipolla, D. and Sokolich, L. (2005) *Utilising a Safety Culture Management Approach in the Australian Construction Industry* [Online]. Available: http://eprints.qut.edu.au/archive/00003797 [20 September 2015].

Bird, S. (2003) Sex composition, masculinity stereotype dissimilarity and the quality of men's workplace social relations. *Gender, Work and Organisation*, **10**(5), 579–604.

Chan, P. and Kaka, A. (2007) The impacts of workforce integration on productivity. In A.R.J. Dainty, S. Green and B. Bagihole (eds), *People and Culture in Construction: A Reader*, pp. 240–54. Taylor & Francis, Oxford.

Chaudhry, S. (2014) *Diversity: Brilliant for the construction industry*. The College of Estate Management, Reading.

Cipolla, D., Sheahan, V.L., Biggs, H. and Dingsdag, D. (2006) *Using Safety Culture to Overcome Market Force Influence on Construction Site*

Safety [Online]. Available: http://eprints.qut.edu.au/3801/1/3801.pdf [20 September 2015].

Clarke, L., Frydendal-Pedersen, E., Michielsens, E., Susman, B. and Wall, C. (Eds) (2004) *Women in Construction*. CLR/Reed Business Information, Belgium.

Cockburn, C. (1983) *Brothers: Male Dominance and Technological Change*. Pluto Press, London.

Cockburn, C. (1991) *In the Way of Women: Men's Resistance to Sex Equality in Organisations*. Macmillan, Basingstoke.

Coffey, M. and Fowler, N. (2010) A comparison of the attitudes of construction workers towards empowerment and participation pre- and post-Egan. In P. Barrett, D. Amaratunga, R. Haigh, K. Keraminiyage and C. Pathirage (eds), *Proceedings of the CIB World Congress – Building a Better World*. CIB, Rotterdam.

Court, G. and Moralee, J. (1995) *Balancing the Building Team: Gender Issues in the Building Professions*. CIOB and Institute for Employment Studies, Brighton.

Dainty, A.R.J., Green, M. and Bagihole, B. (2007) People and culture in construction: contexts and challenges. In A.R.J. Dainty, S. Green and B. Bagihole (eds), *People and Culture in Construction: A Reader*, pp. 3–25. Taylor & Francis, Oxford.

Donaghy, R. (2009) *One Death is Too Many – Inquiry into the Underlying Causes of Construction Fatal Accidents*. The Stationery Office, Norwich.

Egan, J. (1998) *Rethinking Construction*. Department of the Environment, Transport and the Regions, London.

Eisenberg, S. (1998) *We'll Call You if We Need You: Experiences of Women Working in Construction*. Cornell University Press, Ithaca, NY.

Fellows, R., Langford, D., Newcombe, R. and Urry, S. (2002) *Construction Management in Practice*. 2nd Edn. Blackwell, Oxford.

Fryer, B., Egbu., C., Ellis, R. and Gorse, C. (2004) *The Practice of Construction Management*, 4th Edn. Blackwell Publishing, Oxford.

Gadd, S. and Collins, A.M. (2002) *Safety Culture: A Review of the Literature*. HSL/2002/25. Health and Safety Laboratory, Sheffield.

Gregory, C. (2006) Among the dockhands. *Men and Masculinities*, **9**(2), 252–60.

Harris, F., McCaffer, R. and Edum-Fotwe, F. (2006) *Modern Construction Management*. 6th Edn. Blackwell Publishing, Oxford.

Haro, E. and Kleiner, B.M. (2008) Macroergonomics as an organising process for systems safety. *Applied Ergonomics*, **39**(4), 450–8.

Health and Safety Executive (2001) *Establishing Effective Communications and Participation in the Construction Sector*. Research Report 391/2001. The Stationery Office, Norwich.

Health and Safety Executive (2003) *Causal Factors in Construction Accidents*. Research Report 156. The Stationery Office, Norwich.

Health and Safety Executive (2006a) *CN0 COSHH Essentials in Construction: Silica* [Online]. Available: http://www.hse.gov.uk/pubns/guidance/cn0.pdf [20 September 2015].

Health and Safety Executive (2006b) *Watch Your Step in the Construction Industry* [Online]. Available: http://www.hse.gov.uk/construction/campaigns/watchyourstep/index.htm [20 September 2015].

Health and Safety Executive (2009a) *Phase 1 Report: Underlying Causes of Construction Fatal Accidents – A Comprehensive Review of Recent Work to Consolidate sAnd Summarise Existing Knowledge.* The Stationery Office, Norwich.

Health and Safety Executive (2009b) *Phase 2 Report: Underlying Causes of Construction Fatal Accidents – Review and Sample Analysis of Recent Construction Fatal Accidents.* The Stationery Office, Norwich.

Health and Safety Executive (2014) *Health and Safety in Construction in Great Britain, 2014* [Online]. Available: http://www.hse.gov.uk/statistics/industry/construction/construction.pdf [1 October 2015].

Jordan, G., Surridge, M., Mahoney, D., Thomas, S. and Jones, B. (2004) *Mucky, Macho World? The Identification, Exploration and Quantification of Barriers to the Participation of Women in the Construction Industries in Staffordshire* [Online]. Available: https://www.yumpu.com/en/document/view/40432144/mucky-macho-world-university-of-wolverhampton [21 September 2015].

Kletz, T. (2012) The history of process safety. *Journal of Loss Prevention in the Process Industries*, **25**(5): 763–5.

Knutt, E. (2009) We spoke to these people for this article on equality. Does your office have this diversity? *Building Magazine*, 6 October.

Lingard, H. and Rowlinson, S. (eds) (2005) *Occupational Health and Safety in Construction Project Management.* Spon Press, London.

Loosemore, M., Dainty, A.R.J. and Lingard, H. (2003) *Human Resource Management in Construction Projects.* Spon Press, London.

Lupton, D. (2003) Pleasure, aggression and fear: the driving experience of young Sydneysiders. In W. Mitchell, R. Bunton and E. Green (eds), *Young People, Risk and Leisure; Constructing Identities in Everyday Life.* Palgrave Macmillan, Basingstoke.

Manu, P., Ankrah, N., Proverbs, D. and Suresh, S. (2010) The contribution of construction project features to accident causation and health and safety risk: a conceptual model. In C. Egbu (ed.), *Proceedings of the 26th Annual ARCOM Conference*, pp. 261–9. Association of Researchers in Construction Management, Leeds.

McMeeken, R. (2015) Crossing the line. *Building Magazine*, 28 August.

Morton, R. and Ross, A. (2008) *Construction UK: Introduction to the Industry*, 2nd Edn. Blackwell Publishing, Oxford.

Ness, K. (2009) Not just about bricks: the invisible building worker. In A.R.J. Dainty (ed.), *Proceedings of the 25th Annual ARCOM Conference*,

Vol. 1, pp. 645–54. Association of Researchers in Construction Management, Nottingham.

Polesie, P. (2010) What do production managers mean by saying 'I appreciate the freedom on the site'? In P. Barrett, D. Amaratunga, R. Haigh, K. Keraminiyage and C. Pathirage (eds), *Proceedings of the CIB World Congress – Building a Better World*. CIB, Rotterdam.

Rawlinson, F. and Farrell, P. (2010) Construction site graffiti: discourse analysis as a window into construction site culture. In C. Egbu (ed.), *Proceedings of the 26th Annual ARCOM Conference*, Vol. 1, pp. 361–70. Association of Researchers in Construction Management, Leeds.

Rooke, J. and Seymour, D. (2002) Ethnography in the construction industry: competing bodies of knowledge in civil engineering. In S. Seymour and R. Fellows (eds), *Perspectives of Culture in Construction*. CIB Publication No 275. CIB, Rotterdam.

Rooke, J., Seymour, D. and Fellows, R. (2004) Planning for claims: an ethnography of industry culture. *Construction Management and Economics*, **22**(6), 644–62.

Sang, K.J.C., Dainty, A.R.J. and Ison, S.G. (2007) Warning: working in construction may be harmful to your psychological well-being! In A.R.J. Dainty, S. Green and B. Bagihole (eds), *People and Culture in Construction: A Reader*, pp. 127–43. Taylor & Francis, Oxford.

Sawacha, E., Naoum, S.G. and Fong, D. (1999) Factors affecting safety performance on construction sites. *International Journal of Project Management*, **17**(5), 309–15.

Seymour, S. and Fellows, R. (eds) (2002) *Perspectives of Culture in Construction*. CIB Publication No 275. CIB, Rotterdam.

Spanswick, J. (2007) As near as dammit. *Building Magazine*. Issue 13.

UCATT (2015) *Women in Construction* [Online]. Available: http://www.ucatt.org.uk/women-construction [1 October 2015].

Watts, J. (2007) Porn, pride and pessimism: experiences of women working in professional construction roles. *Work, Employment and Society*, **21**(2), 299–316.

World Health Organisation (2002) *Gender and Road Traffic Injuries*. Department of Gender and Women's Health, World Health Organisation, Geneva.

Chapter Three
Safety and Society

... a combination of opportunism, fear, a lot of back-covering and a fair amount of stupidity.

Brown and Hanlon 2014: 123

Many books exploring safety management or safety within a specific industry often ignore the fact that safety does not exist in a vacuum. Whilst the previous chapter exploring the construction site context sets the immediate scene for operational safety management within our industry, what it does not do is look at or even acknowledge what safety *is* within the wider social worlds in which the people who work on sites – from the project manager to the site-cabin cleaner – also live in. Contrary to what some of the ideas of organisational culture may imply, when we go to work we don't instantly change at the front door of the head office or at the site gate into a 'worker', filled with the organisation's values, policies and ideals. We take ideas and understandings from our social lives with us – and these inevitably include ideas about safety.

Unfortunately, within the UK, safety is something that is readily derided, dismissed and pretty much despised. The quotation introducing this chapter is from the book *In the Interests of Safety: The Absurd Rules That Blight Our Lives and How We Can Change Them* and describes the combination of factors that led to one particular historic safety 'scare' in the UK. Whilst this particulate quote is actually a relatively fair assessment of the situation being discussed,

Unpacking Construction Site Safety, First Edition. Dr Fred Sherratt.
© 2016 John & Wiley Sons, Ltd. Published 2016 by John & Wiley Sons, Ltd.

it also helpfully lists some of the key associations safety has within our contemporary society – safety is seen as something enforced by jobsworths, exploited by no-win-no-fee lawyers, is about pre-empting problems around liability and insurance, passing on responsibility and as a result in many cases it becomes something irrational, inconvenient and even just plain stupid.

In fact, the very existence of Brown and Hanlon's book, and its title, also provides a clear indication of how and where safety ranks in UK society. And it's certainly worth noting that this is not one of the sensationalist, intentionally derogatory books about this par-ticular subject – an internet search will quickly find you plenty of those – but instead it raises some very good points about how safety works in the world we currently live in.

And of course this is very relevant to the safety we are looking to explore in this book, the safety found on construction sites. A better understanding of the current 'manifestation' of safety in contempo-rary society is required to, in turn, better understand how people on sites create their own versions of safety that they bring with them to work each day. Whilst there is certainly not scope for a full and detailed exploration of the 'safety of society' (that would be a whole other book – and for each country, safety and society will of course be different – a UK perspective is presented here, although these ideas will certainly have resonance in other countries too), three of the key aspects that hold influence in UK society will be briefly explored and unpacked: the safety of the media, the safety of legislation and the safety that has grown through the (mis-)interpre-tation and enforcement of this law. Safety and society is something that unites and involves people as diverse as celebrities, politicians and bored construction workers, it is something that shapes our society in all aspects from work to play, and it is therefore some-thing that can have considerable influence on safety on sites.

The Media and its Myths

A significant contributor to the safety of society is the media. However, something that is often forgotten about both print and online media is that it is actually in the business of selling advertis-ing, not necessarily informing us of facts, and is well able to utilise the power of indignation to do so; reading how another village fete was ruined by 'health and safety' banning the bunting (untrue – see Health and Safety Executive 2010) provides us with a very satisfying grumble over our morning coffee, but little desire to seek the apology or retraction in the next day's edition. Some, such as Richard Littlejohn, have made it part of their standard remit to deride safety in society. His references to a 'high-viz jacket culture of risk aversion to the point of mental illness' (Littlejohn 2011) are well calculated to bring a snort of derision and nod of agreement whenever 'elf'n'safety' gets a mention in his column. And Littlejohn is not alone, there is even a recognised 'Jeremy Clarkson effect', described as creating an 'unhealthy disrespect of health and safety' through his frequent derogatory comments about safety on his TV shows. With such 'celebrities' adding to the safety of society, their influence has the potential to carry far beyond the immediacy of a TV show into the national psyche.

Yet whilst (at least some of) the media can perhaps be sympathised with to some extent – as Townsend (2013: 55) notes that they do need something to report and if there is no 'juicy disaster' that day they have to 'do the best they can with goggles for conkers and cancelled pancake race stories' – these myths do endure, to the detriment of safety in society. In my lectures to undergraduates on health and safety there is always one student at least who remembers the goggles and conkers story, and I am always happy to set them straight – this wasn't a health and safety rule, this was an overzealous but well-meaning teacher trying to stop kids getting bits of conker in their eye, nothing to do with rules, regulations or the law (see Health and Safety Executive 2007). In his report to the UK government, Professor Löfstedt (2011: 93) specifically highlighted these myths as an inherently problematic aspect of safety in UK society, safety seen as the main reason for 'curtailing beneficial public activities'. Although some within the media have more recently started to share stories of health and safety 'myth-busting', they are still published alongside those who continue to perpetuate the myths, and so the dominance of 'health and safety' as a killjoy and an interfering busybody unfortunately endures.

But this media profile has led to a general unquestioned acceptance in society that 'health and safety' actually *is* the real cause of many management practices to limit participation, enjoyment or fun. But as the Health and Safety Executive's (2014a) Myth Busters Challenge Panel constantly has to point out, using 'health and safety' as a catch-all to deny or ban activities that often have other motivations is highly misleading. For example, stopping students throwing their mortar boards on graduation day cannot be done, as far as the Health and Safety Executive is concerned, simply because of health and safety reasons. Far more likely the motivation for this ban came from the resulting damage to the mortar boards when they land in the melee, rather than any damage to students, and subsequent costs of repair and cleaning to the hire company (Health and Safety Executive 2014b). To label this a health and safety 'concern' is quite simply subterfuge, as the Health and Safety Executive clearly point out, but it is often accepted unchallenged by the majority, drawing on the commonly held misunderstandings of safety in practice.

Perhaps what is wrong here is the lack of challenge such stories and myths receive from society, something at the core of Brown and Hanlon's (2014) book. Safety has become something sacrosanct – who would *want* people to be unsafe or have an accident? – and therefore rules and regulations made in the name of safety become unchallengeable on moral grounds (Townsend 2013). As Starr (1969: 1234) identified in his research around safety back in the 1960s, 'the public generally assumes that the decision making process is based on a rational analysis of social benefit and social risk', however, as Brown and Hanlon (2014) readily demonstrate, this is certainly no longer the case. But of course the result of this is that some of the unchallenged rules and regulations made in the 'interests of safety' are, to put it bluntly, stupid; they have no grounds in rational analysis, and therefore create a frustrating network of restrictions to our everyday lives. Yet this potentially tars *all* safety rules and regulations with the same brush – rules in place on construction sites to protect those working at height will not go untainted by rules made by those keen to pre-empt potential problems that are highly unlikely to ever happen, or unaffected by the many myths that provide some within the media such appetising and outrage-generating content.

However, a media outlet closer to home and one which presents a very different version of safety – one that is not mythical at all – is the construction industry affiliated website *On The Tools TV* (2015), which aims 'to share "building site banter" from Britain and beyond through funny videos and photos' as well as more lofty desires to create a subcontractor networking platform. However, whether those subscribing would want any of the 'actors' in these films working on their sites remains to be seen. For *On The Tools*, safety (or rather the lack of safety) is seen as something funny, something to laugh at, and many of the 'pranks' and 'daily happenings from site' could easily be more formally referred to as serious safety incidents – including collapsing scaffolds or people falling through ceilings. Whilst this site does not (yet) have millions of subscribers, it does present another aspect of the media, one specifically focused on the construction workforce, and one with its own dismissive ideas of safety that can all too readily be shared in site canteens, making this comedic and derogatory version of safety an integral component of the safety of sites themselves.

The growth of the internet, and the resultant seemingly endless variety of 'news' sources means the media are incredibly influential in the way safety works in society. From the familiar erosion of safety found in the *MailOnline* (where safety is positioned as interfering and ridiculous) to the presentation of serious safety incidents as humour through *On The Tools* (where safety is positioned as an irrelevant consideration for the 'unbreakable' construction worker, who is always up for a 'prank'), safety readily permeates our society, and therefore our understandings of both what safety is and how it works. That much media content is inaccurate, belittling and dismissive is not unsurprising – readers are not always that concerned with such details, and the need for sensationalism often wins out when there is website traffic to maintain – but it must be appreciated that the appealing click-bait of another 'crazy' safety story also has influence beyond that day's hit-rate, and has certainly contributed to the ongoing erosion of the standing of safety in today's UK society, and therefore also taints the safety of construction sites.

(Mis-)Interpreting the Legislation

Although the media are often more than willing to tell horror stories about safety 'gone mad', something equally misunderstood and indeed misinterpreted, often through that distorted lens of media mythology, is the framework of safety legislation itself.

Current occupational regulation in the UK is grounded in the Health and Safety at Work etc. Act 1974. In his work that formed the basis of this Act, Lord Robbens set out to create a piece of legislation that was flexible, that made health and safety at work the responsibility of employers *and* employees alike, and tried to ensure appropriate management systems could be developed to manage the specific risks of an industry's operations effectively. When you read the legislation itself, you can actually find very little to argue with in terms of intention and recommended practice. It can easily be found online through www.legislation.gov.uk if you want to see for yourself.

Under this Act, a raft of supporting Regulations have been developed over time. For example, to legislate for general management practices and the need for risk assessment as contained within the Management of Health and Safety at Work Regulations 1999, the need for hard hats on site as set down in the Construction (Head Protection) Regulations 1989, or how to manage safety with reference to the whole life cycle of a project, as prescribed in the Construction (Design and Management) Regulations 2015. However, this wider framework of legislative control has received criticism; it is highly extensive and in some cases complex, although it is often supported by clear guidance from the Health and Safety Executive in many instances, it has generally been seen to have become a rather tangled and knotty ball of 'red tape' (Eaves 2014).

But whilst we are often found bemoaning and grumbling about safety and red tape (risk assessments, for example, are notoriously seen as a bureaucratic bind – usually because they are not carried out correctly, something explored further in Chapter 5), it is worth acknowledging the developmental roots of the legislation we currently have in place, because what is missing from our contemporary social awareness of safety is that the current (actual as opposed to the mythical) occupational safety regulations are the result of long and hard fought campaigns.

Our contemporary UK legislation developed from the earliest Factory Acts of the Victorian age, and many people had to die or sustain serious injuries to bring about any legislative intervention in the first place. The debates from the Houses of Commons and Lords of these times, when the legislation was being proposed and discussed, make for very interesting reading as the familiar arguments around production (allegedly the arch rival of safety legislation), investment, responsibility and liability first emerge.

Indeed, the way these debates still resonate today is something that will be explored later in this book, but there is one key difference between then and now that can be readily identified – and that is the fact that the voices have changed. Today it is also the working people who are bemoaning safety and safety regulation, instead of just the factory owners and businessmen of the day (who were of course also the MPs of the time). As noted by union health and safety representative Dave Putson (2013: 171), 'today some workers seem to take rather casually hard fought and hard won legislation … with little awareness of how long it took to secure health provision for all'. The journey (some might call it a struggle) of safety has seemingly been forgotten, a societal carelessness that has likely contributed to its contemporary repositioning – from something to be cherished and valued in the workplace, to something to be dismissed and derided. It is ironic that the arguments against legislated safety, first made by the top-hatted factory owners and businessmen of Victorian times, are so willingly taken up and repeated by those wearing hard hats and rigger boots today.

And these voices were joined in 2012 by UK Prime Minister David Cameron, on his own mission to 'tackle' safety legislation. Cameron stated his belief that 'health and safety legislation has become an albatross around the neck of British business' (Woodcock *et al.* 2012) and launched the 'Red Tape Challenge' (2015) to reduce the amount of regulation and bureaucracy in all aspects of life in the UK.

In making this statement, Cameron was aligning his government to the popular ideas of safety; that health and safety, or to use more derogatory Littlejohn-esque shorthand, 'elf'n'safety', in the UK has 'gone mad'. He described the health and safety culture as a 'monster', which Townsend (2013: 45) rightly notes can be a very useful thing 'it helps to have something evil. You know where you are with evil … you can fight it'. But of course this also has consequences for safety and society; if the representative of the UK people is happy to criticise and demonise safety legislation in its current incarnation, and add his voice to the derision of safety in our society, who is to challenge it?

Well, unfortunately for Cameron, the man he tasked to review health and safety in the UK did not entirely agree with him. Professor Löfstedt's report, *Reclaiming Health and Safety For All: An Independent Review of Health and Safety Legislation*, in fact concluded that:

> … in general, there is no case for radically altering current health and safety legislation. The regulations place responsibilities primarily on those who create the risks, recognising that they are best placed to decide how to control them and allowing them to do so in a proportionate manner. There is a view across the board that the existing regulatory requirements are broadly right, and that regulation has a role to play in preventing injury and ill health in the workplace. Indeed, there is evidence to suggest that proportionate risk management can make good business sense. (Löfstedt 2011: 1)

Away from the political hyperbole around safety legislation, it can readily be argued that to dismiss what the Health and Safety at Work etc. Act 1974 actually set out to do seems very much a case of throwing the baby out with the bathwater, when its basic goals of protection from harm are rationally considered. Indeed, Putson (2013: 175) goes further to suggest that, "The repeal of health and safety legislation, undermining or neutering it ... not only constitutes an extreme act of vandalism, it is also intellectually incoherent'. And it would seem that the government's own advisor would be inclined to agree with him. It could even be suggested that the red tape is not actually constrictive, rather it is simply holding everything in.

But the 'Red Tape Challenge' (2015) appears to be an ongoing project, proudly boasting on an infographic: '84%: the proportion of health and safety legislation being scrapped or improved', although details of what this precisely involves are rather harder to find. Professor Löfstedt's report is rather less publicised, there are no infographics with nice big percentages on them. As a result, the dominant political discourse remains that safety legislation is still a 'monster', mainly composed of red tape, which needs to be slain – an interesting reframing of the old Victorian mill-owners' arguments for reductions in control and legislation. Well, perhaps in the UK our MPs have not changed that much after all.

However, key to Löfstedt's assessment of UK legislation was the fact that the 'problem lies less with the regulations themselves and more with the way they are interpreted and applied' (Löfstedt 2011: 2). And this is critical. But the interpretation of legislation is of course made from within the social context of safety, and if this is grounded in myths, derision and misunderstandings, then legislation can naturally become a tool for the worse, as has become apparent in the growth of safety litigation in the UK.

Where there's blame …

One of my clearest memories from my time as a site manager was of an accident. Someone had chosen to jump over some stacked materials to take a shortcut, rather than using the clear walking route that went around them, landed badly and hurt his ankle. But despite the circumstances, and the fact that I knew I couldn't have done more myself to prevent this happening, as I helped him off the site to a seat in the canteen the call came loud and clear from the scaffolding above:

Where there's blame there's a claim!

And apparently there was. I prepared a file full of information, records, risk assessments and method statements, I even got in trouble for my photos of the area where the incident occurred – showing so tidy a site that it didn't look like any work was being done! But he still got his claim; it was simply easier and more economically sensible to pay out than pay to fight.

And this is a clear example of what has been described as the 'compensation culture' of the UK. Liability and responsibility could have of course been debated in this particular case, yet the claim was made, and indeed paid. This change in how we see and respond to safety incidents is a result of the hijacking of safety by ambulance-chasing lawyers, enabled by the Access to Justice Act 1999 and its conditional fee agreements (in more common terms, no win no fee) which reduced the financial risks of claiming in marginal circumstances, and arguably planted and nurtured such a 'compensation culture'. Concerns of liability and litigation abound, although not as a direct consequence of health and safety law (the difference between civil litigation under tort law and prosecution under criminal law often causes confusion), but rather as a result of a now unimpeded, zealous and well-paid legal community, and those awarding compensation where common sense may suggest it is not always deserved. As Brown and Hanlon (2014: 17) suggest 'rarely has a single law change had such a pernicious effect on the national psyche.

The existence of a 'compensation culture' has long been the subject of debate; despite the Better Regulation Task Force (BRTF) stating in its report of 2004 that a 'compensation culture' simply did not exist in the UK, others suggest the information was not available to make such statements. Instead there was evidence that 'some sorts of accident claims have risen (from a relatively low base) and that the overall costs of personal injury settlements have gone up' (Williams 2006: 513), and ideas of a compensation culture remain a prominent aspect of safety in society.

It is therefore unsurprising that a vow to 'kill off the health and safety [compensation] culture for good' (Woodcock et al. 2012) was also included within Prime Minister David Cameron's safety remit. He tasked Lord Young to review the situation, and in his foreword to the report eventually produced Cameron (2010) stated that the need for this review was because:

> The standing of health and safety in the eyes of the public has never been lower.

Yet the reasons for this may not be as clear as Cameron suggests, a thread of his argument drawing on the fact that:

> Newspapers report ever more absurd examples of senseless bureaucracy

That the newspapers are not necessarily telling the truth (see Health and Safety Executive myths ad infinitum!) does not seem to be relevant, and ignores the fact that the integrity of the British press is in fact an entirely different problem from health and safety. With regard to occupational health and safety, Cameron goes on to boldly claim that:

> We're going to end the unnecessary bureaucracy ... from our businesses.

Which was actually something that Professor Löfstedt was unable to help him with when the time came around to do so. But yet again, this clear positioning of safety in the negative by the UK Prime Minister clearly illustrates its place in UK society, and serves to perpetuate the rhetoric that surrounds safety as something to be derided.

Young's (2010: 7) report *Common Sense Common Safety* did find evidence of a 'compensation culture' driven by litigation at the heart of the problems that so beset health and safety today. He identified a 'climate of fear' caused by 'an overzealous approach to applying the H&S regulations ... exacerbated by insurance companies' and compounded by no-win-no-fee lawyers. These findings were later echoed by Professor Löfstedt (2011: 6), who linked the misinterpretation of legislation directly to this context, stating that 'the threat of being sued can be a key driver for duty holders going beyond what the regulations require'.

It is this 'compensation culture' that forms the third aspect of Brown and Hanlon's (2014: 3) 'unholy alliance of official self-importance, media hysteria and commercial exploitation'. The commercial opportunities for both consultants and lawyers are seemingly just too tempting to prise them away from safety, and the UK government's attempted intervention in the form of the Compensation Act of 2006 has actually done little to reduce the flow of cases. As a result safety has become enmeshed in a mire of litigation, companies drowning in the *evidence* of responsibility (not the responsibility itself) and the ongoing quest to shift liability – all of which are of course distractions from the actual business of keeping people safe at work.

Safety and Society and Construction Sites

A thread that runs throughout government reports and the more rational contemporary critiques of safety and society is that the perception people have about safety is critical. Both Lord Young and Professor Löfstedt agree on this, and Löfstedt (2011: 87) specifically highlights the media as a driving force in the creation and perpetuation of the common misunderstandings around safety.

But unfortunately it is only really the perception that matters; it is perception that makes people overcompensate and exceed the actual requirements of legislation, it is perception that creates guidance that doesn't really work in practice, it is perception that makes people produce piles and piles of paperwork just in case they are sued in the future, it is perception that gives safety its low standing in contemporary UK society. It is this same perception that means people don't take anything about safety seriously, it is also perception that means people don't follow the rules, it is perception that makes people think safety does not really apply to them, and unfortunately it is perception that leads to accidents and incidents.

And these wider social ideas and understandings of safety undoubtedly influence that found on construction sites. As a result, when the question 'what is safety?' is asked, immediate responses are often negative and frequently fail to get beyond paperwork, bureaucracy and 'arse-covering', rarely reaching the arguably far more relevant aspects of freedom from harm, accidents or incidents. The context has influenced the concept; safety has become a derisible aside, a negative influence on our daily lives and a management and administrative burden on our work. Although this isn't *actually* safety, it might well as well be; it is what safety seems to have become.

This brief overview has been considered here in order to illustrate the importance of *how* we talk about safety. Our contemporary understandings of safety have developed and crystallised through these dominant channels – the way we *perceive* safety to work is highly influential, and often more so than the mundane realities of how it *actually* works – as well as other more specific and individual influences, and has consequently permeated through society and into its workplaces. The context of safety on construction sites is therefore made up of much more than the industry structure, ways of working and the site rules as implemented by the site management; this wider social context must also be considered and indeed goes some way to explain why safety is such a problem on UK sites.

Understanding People

As we have seen above, in order to better understand how safety works the many complex influences at play must be acknowledged and, if Townsend (2013: 133) is correct in making the statement that 'there is no understanding common to all members of society of what makes health and safety work', this will not be a simple task. This chapter has already shown that safety is a social thing; and what safety *is* will be influenced by many different social aspects including our daily work practices, the media, and conversations with family or with friends down the pub. All of these different ideas come together to produce an understanding of safety, which will in turn affect how we use it in our working lives. Do we take the risk assessment seriously and carefully think through each step and seek out new methods to make the work safer? Or do we cut and paste the generic risk assessment carried out in 1998 onto a new page and ask our team to sign up to it without anyone even reading it through? But as this simplistic example is still able to readily demonstrate, the key to safety is people and how people create safety in practice, and therefore key to an understanding of safety is an understanding of people and their contexts, be they focused on the confines of their workplace or taken from the perspectives of their wider society as a whole.

Many of the ways we try to understand people use ideas and knowledge originally developed to understand things that are far less complicated. In science, knowledge comes from experiments and calculations, which can then be used to develop universal theories. For example Newton's third law of motion (for every action there is an equal and opposite reaction) was put down in the seventeenth century and is still employed in the structural analysis of the design of buildings today. Such laws and rules govern how things should and will behave under certain circumstances, and therefore produce universal 'truths'. They are based on the belief that the world *can* be tested and experimented upon and knowledge is seen as fact, based on empirical observations, using scientific methods of control, standardisation and objectivity to establish its theories (Henn *et al.* 2006).

But people are very different to the steel beams and concrete of science, and when we try to measure or quantify them as if they were inert subjects in an experiment, problems quickly arise. People can be inconsistent and variable, they can be completely illogical (as Spock frequently had to point out on the Starship Enterprise), they change their minds and they have different responses depending whether they had a nice lunch, whether they are off on holiday tomorrow, or if they forgot their umbrella in the rain and are currently sitting there grumbling in damp and uncomfortable clothes.

As a result of this inconsistency there have been various developments in the social sciences which seek to explain, allow for, and ultimately acknowledge the awkwardness of people. Theories have been put forward which try to create 'rules' and 'truths' about people and how they behave, yet be flexible enough to allow for such inconsistencies. Although the names might be new, many of these ideas are very familiar to practice, including the safety management practices of the construction industry. Various theories have been used to develop various safety management programmes, based on the specific ideas about people and how best to influence them. Three of these theories, and their relationships to safety management on construction sites, are briefly examined here.

Behaviourism

Behaviourism can be seen as the 'scientific' way to research people; behaviours can be described scientifically without any recourse to internal physiological events or hypothetical constructs, ideas such as 'the mind'. Through experimental behavioural analysis of rats and pigeons, B.F. Skinner (1978) developed his theories regarding the impact and role of the environment in the determination of human behaviour. His theory of operant conditioning argues that behaviour comes under the control of stimuli, and that cognitive thought processes such as 'intention' and 'purpose' were merely invented explanations for the repeating of previously rewarded behaviour.

Although the construction workforce is not made up of rats and pigeons, behaviourism can be recognised as the underlying theory of many different safety programmes found on construction sites, for example the Goals and Feedback approach (IOSH 2006) is based on behaviour modification. Drawing on Skinner's work, this approach assumes people will repeat behaviours that have a favourable consequence, and tend not to repeat behaviours that bring unfavourable consequences (David and Newstrom 1989; Rachlin 1991) such as punishment or discipline. So, if safe working receives a reward in terms of financial benefit or prestige, then safer working should be the norm throughout the project. But studies have shown that if an unsafe act has positive consequences, such as getting the job done more quickly and rarely causes an accident, then this act is likely to continue (Saari 1994), despite the interventions of management. Other limitations have been found when trying to change behaviours that operatives find disagreeable, such as having to wear inconvenient and uncomfortable personal protective equipment (PPE) (Cameron and Duff 2007). And despite positive rewards and the presence of punishments, people still don't always comply with safety rules and regulations, something that still rings true on sites today. Behaviour modification programmes were taken up with some enthusiasm in the construction industry in the 1980s, but have since declined in popularity for a number of reasons, not least their limited scope for effectiveness.

Cognitive theories

Cognitive theories broadly adhere to the idea that each of us has a distinct and tangible 'mind', the internal 'cognitive machinery' that drives human understanding and experience. Cognitive theories have developed 'truths' about many different things, for example ideas around motivation (Hale 2008; Ridley and Channing 2008), cognitive dissonance (Baron *et al.* 2006) or the determinants to planned behaviour (Ajzen 2005). However, these theories often compete, overlap and conflict with each other, and can come with caveats stating the required context for the theory to hold (Farrell 2011).

An example can be seen in expected utility theory, which states that people are rational decision makers that will weigh up the utilities of outcomes through probability, and seek to maximise expected utility from their actions (Baron 2008; Hardman 2009). This goes some way to explain why people will still take risks on site – if it will get the job done faster, the balance of safety versus productivity comes into play. However experimental research has disproved this basic assumption within certain contexts – people do not make rational decisions, and consequently further explanations have been developed, such as the Allais paradox, which states that people do not always choose decision options that maximise expected utility (Hardman 2009), reflecting that human behaviour is neither consistent (Ajzen 2005) nor rational (Perrow 1999).

Such variability in people has been explored through approaches such as the systems model of human behaviour, which acknowledges conflict between motivations and goals, including conflicts between the long and short term (Ridley and Channing 2008), again looking to cognitive explanations for underlying theories of decision making. The way people organise and process their thoughts has been broken down into construals and schemas, employed to gain accurate understandings and apply knowledge to new situations (Aronson *et al.* 2007). Heuristics are ascribed to people as the mental shortcuts they use to reduce the complexity of everyday judgements (Strauch 2004). They are employed to create the construct of 'bounded rationality' in people (Hardman 2009), to explain why people are not absolutely rational in their behaviour. The employment of these heuristic shortcuts is seen to bypass rationality, for example the 'availability heuristic' means people will judge a situation in terms of the most readily available information. An example put forward by Perrow (1999) states that a recent publicised aeroplane crash will increase fear of flying due to focus on that particular piece of evidence, rather than the thousands of successful flights undertaken every day. There are also other tendencies ascribed to people that exercise influence on behaviour, such as the optimism bias, the predisposition to expect that things will turn out well, the overconfidence barrier, which places greater confidence in personal judgement than is justified (Hale 2008), and the planning fallacy, the tendency to make optimistic predictions about how long a task will take (Baron *et al.* 2006) – something very familiar to anyone who has worked in construction!

Other key ideas within the cognitive theories that are often employed within social research are those of values, attitudes and beliefs (Baron *et al.* 2006). These elements are often seen as the basic criteria of many social phenomena, including the highly complex concept of culture, although it is attitudes, the inherent disposition to respond favourably or unfavourably to an object/person/event (Aronson *et al.* 2007), that are most frequently used, due to their accessibility through tools such as questionnaires or observed behaviours (Ajzen 2005). In construction, safety management often draws on this way of thinking in the use of safety climate surveys, completed by the workforce to gain some measure of safety or more specifically the 'safety culture' of a project or organisation (Guldenmund 2007; Glendon 2008). Such surveys are common practice on large sites or within large organisations, despite challenges made to the relevance, accuracy and indeed use of such data in terms of its reflection of 'safety culture' once it has been collected (Hopkins 2002).

Safety interventions based on cognitive theories have been readily employed on construction sites. Focused applications of cognitive ideas include those put forward by Gheisari et al. (2010) around the concept of situation awareness, and the application of goal-directed cognitive task analysis techniques to identify safety critical information and the requirements for decision making on sites by safety managers. Whilst they suggest this approach has 'great potential of improving safety management practices on jobsites' (ibid. 2010: 317), they also acknowledge that one of the challenges in such a cognitive approach is that 'situational awareness may vary from individual to individual' (ibid. 2010: 318), and as a consequence there are difficulties in developing a unified application to practice. More holistic applications of cognitive thinking can be seen in the 'cultural change' programmes first implemented in the UK on sites in the late 1990s (Health and Safety Executive 2008). One such example is the Incident and Injury Free programme, originally developed in the USA, which appeals to the rational self to 'make safety personal' and influence behaviour by reminding operatives that they are also husbands, fathers and sons – drawing heavily on ideas of cognitive dissonance, which suggests that the workforce will struggle to place this family identity within a workplace of risk taking and unsafe actions, and the workers will therefore adjust their behaviour to suit that of the family man, and change their overall attitudes towards safety as a result. However, research has shown that there is no direct evidence of success for this approach, or changes in behaviour as a result of these methods (Health and Safety Executive 2008) within the highly complex construction site environment.

Social constructionist theory

Social constructionist theory examines individuals simply in terms of their social context and interactions with others (Wiggins and Potter 2007) – which actually means it does not really examine individuals at all. Instead, this theory seeks to explore the social world not as it 'really' is – as we think of the 'truths' that we can establish around steel beams and concrete – but as a place constructed by people through their constant social interactions (Burr 2003). All social interactions and shared practices result in shared versions of knowledge within particular communities (Gergen and Gergen 2003) – essentially this knowledge becomes 'constructed'. And as a result 'truth' simply becomes the *current* accepted way of understanding the world (Burr 2003), depending who we are with and what we are doing.

Because people are constantly moving between different environments and different interactions, the language and practices they employ to perform different actions within these different social contexts (Potter *et al.* 2007) inevitably results in frequent shifts in self-presentation and identity (Augoustinos *et al.* 2006). The assumption is made that identity is also rather incoherent, fleeting and fragmented (Edley 2001), and there is no distinct and rigid 'mind' to get hold of. Change is therefore the only constant in both people and the social worlds they construct (Gergen 2009).

As a result of this, variation within the way people construct their shared understandings of their social worlds is accepted and even expected (Wetherell and Potter 1992), and made without the need for explanatory devices; the very nature of the approach assumes inconsistencies as people do different things within different social contexts (Alvesson and Sköldberg 2000; Augoustinos *et al.* 2006; Billig 2007). Indeed, people can be found to offer different evaluations of the same thing on different occasions, and can even change and shift in their ideas within different parts of the same conversation (Potter 1998).

The aim of a social constructionist approach is not to seek out and identify any underlying attitudes as 'facts', but instead to explore the practices through which different shared knowledge is constructed and legitimised, with no assumptions of consistency. This means that social constructionism results in an examination of people that cannot establish identity, or attitudes, or predict behaviours, as these constructs no longer apply; indeed there is no 'truth' to be established. However, it can be used to explore shared *understandings* within certain contexts. Rather than focus on individuals it looks at the wider social network and so the concept of truth is not excluded, rather the difference is in the perspective taken; through the use of language truths become local, negotiated understandings that are produced in social life rather than objective principles that direct the way social life develops (Augoustinos *et al.* 2006).

This approach might seem rather fanciful and unhelpful – surely it would be better to pin down *why* people do things, and seek explanations? – but to see how this approach works in practice, consider the following example of a roofer on site.

An example from construction site life

A male roofer sits in the induction room and listens carefully, he nods in all the right places, agrees with the site rules and the fundamentals of the Incident and Injury Free (IIF) safety programme in place on site, and signs up to his method statement and risk assessments, which clearly state he will use the lanyard and harness at all times when working on the roof. A mere two hours later he is seen working on the pitch of a wet metal roof with his lanyard still attached to his harness and not to the safe anchor point a few feet away. This is an unsafe behaviour which could result in a serious, potentially fatal accident, should he lose balance, slip and fall.

So why on earth would he do this?

From a behavioural perspective, the roofer is at the mercy of external contingencies (Skinner 1978), the need for productivity and progress towards the clients' deadlines are clear goals which must be achieved (Health and Safety Executive 2003; Rawlinson and Farrell 2008; Health and Safety Executive 2009) as well as more personal goals set for workers paid on 'price', where the daily output equals the daily pay. The behaviourist approach looks for, and in this case can easily find, external contingencies which provide 'explanations' for the roofer's behaviour. But this tells us nothing new. These aspects of construction site life have been researched many times before, including with reference to their potential impact on safety (see for example Langford *et al.* 2000; Health and Safety Executive 2003; Cipolla *et al.* 2006; Choudhry and Fang 2008; Rawlinson and Farrell 2008; Donaghy 2009). To investigate this incident from a behavioural perspective would once again provide objective 'reasons' for the roofer's behaviour, without any further developments in terms of understanding.

From a cognitive perspective, the roofer's behaviours can be explained through a variety of constructs to try to clarify the machinery driving his mind, and therefore making him act unsafely. It can be suggested that the roofer has employed heuristics in his approach to work, as these mental shortcuts are exactly the kind of rough judgements that will be applied in such a familiar and every-day setting (Hardman 2009). For example, the representativeness heuristic will have classified his situation according to how similar it is to a typical case, and then adjusted his attention levels accordingly (Aronson *et al.* 2007); he has fitted this type of roof a hundred times before and so he can work on automatic pilot. However, within complex environments which require complete accuracy for safe performance, the use of heuristics can prove problematic; whilst the social world usually allows for rough judgements when the cost of inaccuracy is not too great, on a construction site where safety is concerned the repercussions can be far more significant.

In addition to heuristics, other cognitive explanations can also be given for his behaviour. For example the roofer could have fallen foul of optimistic bias, and simply be sure that this time he will be alright, or become directed by the overconfidence barrier, and placed greater confidence in his personal judgement of safety than is justified (Baron *et al.* 2006; Hardman 2009). Also, the roofer is likely to have been influenced by other, more general circumstances that morning; with his reactions potentially coloured by past experiences (he's never had an accident … touch wood), his state of mind at the time (row with the missus before he left for work), his well-being and health (sore leg from Sunday league football, or sore head from the post-match celebrations) and even what type of day he is having (raining again!) could have influenced his behaviour. With his short-term goal (to make his money quickly so he can get home in time to start tea for his wife to say sorry) in likely conflict with his long-term goal of self-preservation, it is possible that the roofer was rushing to carry out his tasks and failing to check his equipment correctly, therefore jeopardising his long-term goal of ensuring his own safety (Hale 2008).

It can clearly be seen that a cognitive approach allows for highly detailed explanations of the roofer's behaviour, drawing on a vast array of mental constructs, influences and causal factors. However, in terms of simplification or understanding of behaviour, this approach is not particularly helpful. Risk-taking behaviour can be explained through a variety of heuristics (Perrow 1999; Aronson *et al.* 2007), supported by the implementation of prospect theory, but this must then be justified by the Allais paradox when the highlighted behaviour does not comply (Hardman 2009). And that is notwithstanding the wider social consideration that safety is a load of rubbish anyway, even the prime minister thinks so .... But to explore this scenario through a cognitive approach is arguably limited to providing explanations for behaviours based on observations or the reliance on language as directly reflective of these different cognitive constructs when the roofer is questioned after the event. However, this may prove problematic; given the subject matter of safety, self-implication (Lee 2000), or rather the avoidance of self-implication, could become critical, in addition to problems with asking people to think about things they would not normally talk about (Inglis 2005). Saying why we did something, especially after the event when we can reflect and evaluate in the cold light of day that it probably wasn't a particularly good idea, can be a very difficult task. In terms of any development of understandings of the roofer's behaviour, cognitive theories are arguably limited by the inherent constraints in place on actually establishing whether the cognitive explanations made are indeed the right ones, and if they can be checked and confirmed with the roofer after the event.

From the social constructionist perspective, the variability in the roofer's behaviour can be much more easily understood. He was merely constructing his own socially accepted version of safety as required at that time and in that particular context. A different understanding of safety is found within the induction room, where he has just been reminded that he is a husband/son/father and safety is his responsibility, than out on the construction site itself, where he is a roofer, and so safety may not be held in such high regard alongside the other contextual influences of productivity and progress. Acknowledging and understanding that there are different ideas of what safety *is* within different contexts already provides us with some new insights that could help us start to tailor safety management practices.

And it must be noted that our roofer is not consciously acting from a manipulative position. Rather his social identities will shift with his immediate environment, creating a variety of different ideas which vary depending on what is most dominant in his work context: safety or productivity. From a social constructionist perspective, the roofer has no stable cognitive self, rather a number of multiple identities (Augoustinos *et al.* 2006): is the person concerned actually a roofer or a father? The answer can easily be both. This provides a different explanation and indeed enables us to understand why attention and support was given towards safety by the self in the induction room, yet disregarded by the self on site, amongst the banter and bravado, to construct safety in that particular context.

It can therefore be seen that a social constructionist approach can give a different perspective of construction site safety and enable us to unpack it using different thinking. Rather than only trying to identify external contingencies that influence context, or segmented explanations that only apply in certain circumstances and not others, a constructionist approach aims to explore *what* is being constructed by these shifting identities in their varied contexts. We are able to ask the questions posed earlier in this introduction. What is safety on site? Do we agree on our definition? How do we talk about it? How is safety associated with practice? Does it work? In our roofer's case the answer to this last question is unfortunately no.

The Rest of This Book

Social constructionism seeks to explore people within different social contexts, how they develop, contribute to and perpetuate different forms of shared knowledge through their interactions, to reveal how ideas are understood and taken for granted in different societies (Berger and Luckmann 1966). In seeking to 'tell the truth' in terms of the conventions of a particular social group (Gergen 1999), this approach is able to reveal the variability, change and inconsistency that is an integral part of our social lives. In this book, the theory is applied to safety – how people understand, know and talk about safety, and how this influences its role within everyday work practices. To ultimately ask what is being constructed about *safety*, alongside the concrete and steel, on large UK construction sites.

This book explores the use of language – either the written word or talk – which forms the foundations of social constructionist theory. Language is how people represent themselves and their worlds as they come together in social interactions; language is how we *construct the world*. However, language cannot be considered to simply represent 'the truth', it requires exploration and analysis to reveal what shared realities, conventions and practices are actually being used by people in their different interactions and contexts. *What* people say is only partly of importance, *how* they say it is of equal merit – What associations do they make? How do they position key ideas?

What justifications or excuses do they make? These are the *ways in which language is used to construct different realities and worlds, the 'set of meanings, metaphors, representations … that in some way together produce a particular version of events'* (Burr 2003: 64), in this case construction site safety.

This explains why people, and the realities in which their shared understandings come together, can be so variable; they use language in different ways in different situations to construct their worlds at different times and in different places (Gergen and Gergen 2003; Augoustinos *et al.* 2006). Our roofer from the previous example can be better understood from the perspective of safety relevant to his particular context; the disassociation of safety from productivity is a dominant way safety is positioned on sites, and one which is readily identifiable here. It is through such analysis that specific aspects of our social environment, such as safety, can be unpacked and revealed within the construction site environment.

Within this book, the quotes and extracts have been taken from real life; from the author's conversations with managers, operatives and foremen about safety, from safety documentation that forms safety management systems, and from the safety signage used on sites. Examples within this text are representative of those used in the wider research that underpins this book, and they are shared here to illustrate some of the common ways in which safety is constructed on sites, as well as providing a good starting point for discussions and explorations of not only what safety *is* but also how it *works* on construction sites.

References

Ajzen, I. (2005) *Attitudes, Personality and Behaviour.* 2nd edn. Open University Press, Buckingham.

Alvesson, M. and Skoldberg, K. (2000) *Reflexive Methodology: New Vistas for Qualitative Research.* Sage, London.

Aronson, E., Wilson, T.D. and Akert, R.M. (2007) *Social Psychology.* 6th edn. Pearson Education, New Jersey.

Augoustinos, M., Walker, I. and Donaghue, N. (2006) *Social Cognition: An Integrated Introduction.* 2nd edn. Sage, London.

Baron, R.A., Byrne, D. and Branscombe, N.R. (2006) *Social Psychology.* 11th edn. Pearson Education, London.

Berger, P. and Luckmann, T. (1966) *The Social Construction of Reality: A Treatise in the Sociology of Knowledge*. Penguin, London.

Billig, M. (2007) The argumentative nature of holding strong views: a case study. In J. Potter (ed.), *Discourse and Psychology Volume 2: Discourse and Social Psychology*. Sage, London.

Brown, T. and Hanlon, M. (2014) *In the Interests of Safety: The Absurd Rules That Blight Our Lives and How We Can Change Them*. Sphere, London.

Burr, V. (2003) *Social Constructionism*. 2nd edn. Routledge, East Sussex.

Cameron, D. (2010) Foreword by the Prime Minister. In D.I. Young (ed.), *Common Sense Common Safety*. Cabinet Office, HM Government, London.

Cameron, I. and Duff, R. (2007) A critical review of safety initiatives using goal setting and feedback. *Construction Management and Economics*, **25**(5), 495–508.

Choudhry, R.M. and Fang, D. (2008) Why operatives engage in unsafe work behaviour: investigating factors on construction sites. *Safety Science*, **46**(4), 566–84.

Cipolla, D., Sheahan, V.L., Biggs, H. and Dingsdag, D. (2006) *Using Safety Culture to Overcome Market Force Influence on Construction Site Safety* [Online]. Available: http://eprints.qut.edu.au/3801/1/3801.pdf [20 September 2015].

David, K. and Newstrom, J.W. (1989) *Human Behaviour at Work: Organisational Behaviour*. 8th edn. McGraw-Hill, Maidenhead.

Donaghy, R. (2009) *One Death is too Many – Inquiry into the Underlying Causes of Construction Fatal Accidents*. The Stationery Office, Norwich.

Eaves, D. (2014) *Two Steps Forward, One Step Back – A Brief History of the Origins, Development and Implementation of Health and Safety Law in the United Kingdom, 1802–2014, History of Occupational Safety and Health* [Online]. Available: http://bit.ly/1nJPvjJ [25 September 2015].

Edley, N. (2001) Analysing masculinity: interpretive repertoires, ideological dilemmas and subject positions. In M. Weatherell, S. Taylor and S.J. Yates (eds), *Discourse as Data: A Guide for Analysis*, pp. 189–228. Sage, London.

Farrell, P. (2011) *Writing a Built Environment Dissertation: Practical Guidance and Examples*. Wiley-Blackwell: Chichester.

Gergen, K.J. (1999) *An Invitation to Social Construction*. Sage, London.

Gergen, K.J. (2009) *An Invitation to Social Construction*. 2nd edn. Sage, London.

Gergen, M. and Gergen, K.J. (2003) *Social Construction: A Reader*. Sage, London.

Gheisari, M., Irizarry, J. and Horn, D. (2010) Situation awareness approach to construction safety management improvement. In C. Egbu (ed.),

Proceedings of the 26th Annual ARCOM Conference, pp. 311–18. Association of Researchers in Construction Management, Leeds.

Glendon, A.I. (2008) Safety culture and safety climate: how far have we come and where could we be heading? *Journal of Occupational Health and Safety Australia and New Zealand*, **24**(3), 249–71.

Guldenmund, F.W. (2007) The use of questionnaires in safety culture research – an evaluation. *Safety Science*, **45**(6) 723–43.

Hale, A. (2008) The individual and safety. In J. Ridley and J. Channing (eds), *Safety at Work*. 7th edn. Butterworth Heinemann, Oxford.

Hardman, D. (2009) *Judgement and Decision Making – Psychological Perspectives*. Blackwell Publishing, Oxford.

Health and Safety Executive (2003) *Causal Factors in Construction Accidents, RR156*. The Stationery Office, Norwich.

Health and Safety Executive (2007) *Myth: Kids Must Wear Goggles to Play Conkers* [Online]. Available: http://www.hse.gov.uk/myth/september. htm [25 September 2015].

Health and Safety Executive (2008) *Behaviour Change and Worker Engagement Practices within the Construction Sector, RR660*. The Stationery Office, Norwich.

Health and Safety Executive (2009) *Underlying Causes of Construction Fatal Accidents – A Comprehensive Review of Recent Work to Consolidate and Summarise Existing Knowledge*. Phase 1 Report. The Stationery Office, Norwich.

Health and Safety Executive (2010) *Myth: Health and Safety Bans Bunting* [Online]. Available: http://www.hse.gov.uk/myth/aug10.htm [25 September 2015].

Health and Safety Executive (2014a) *Myth Busters Challenge Panel Findings* [Online]. Available: http://www.hse.gov.uk/myth/myth-busting/ index.htm [25 September 2015].

Health and Safety Executive (2014b) *Case 309 – University Bans the Throwing of Mortar Boards on Graduation Day* [Online]. Available: http://www.hse.gov.uk/myth/myth-busting/2014/case309-university-bans-throwing-of-mortar-boards.htm [25 September 2015].

Henn, M., Weinstein, M. and Foard, N. (2006) *A Short Introduction to Social Research*. Sage, London.

Hopkins, A. (2002) *Safety Culture, Mindfulness and Safe Behaviour: Converging Ideas?*, Working Paper 7, National Research Centre for OHS Regulations, The Australian National University.

Inglis, D. (2005) *Culture and Everyday Life*. Routledge, Oxford.

IOSH (2006) *Behavioural Safety – Kicking Bad Habits*. Direction 06.1. Institution of Occupational Safety and Health, Leicestershire.

Langford, D., Rowlinson, S. and Sawacha, E. (2000) Safety behaviour and safety management: its influence on the attitudes of workers in the UK construction industry. *Engineering, Construction and Architectural Management*, **7**(2), 133–40.

Lee, R.M. (2000) *Unobtrusive Methods in Social Research*. Open University Press, Buckingham.

Littlejohn, R. (2011) *This elf'n'safety madness is par for the course...* [Online]. Available: http://www.dailymail.co.uk/debate/article-2064445/Health-safety-regulations-This-madness-par-course.html [25 September 2015].

Löfstedt, R.E. (2011) *Reclaiming Health and Safety For All: An Independent Review of Health and Safety Legislation*. Department for Work and Pensions, UK Government, London.

On The Tools (2015) On The Tools TV [Online]. Available: http://onthetools.tv/ [25 September 2015].

Perrow, C. (1999) *Normal Accidents – Living with High Risk Technologies*. Princeton University Press, New Jersey.

Potter (1998) Discursive social psychology: from attitudes to evaluative practices. *European Review of Social Psychology*, **9**(1), 233–66.

Potter, J., Weatherell, M., Gill, R. and Edwards, D. (2007) Discourse: noun, verb or social practice? In J. Potter (ed.), *Discourse and Psychology Volume 1: Theory and Method*. Sage, London.

Putson, D. (2013) *Safe at Work? Ramazzini versus the Attack on Health and Safety*. Spokesman Books, Nottingham.

Rachlin, H. (1991) *Introduction to Modern Behaviourism*. 3rd edn. W.H. Freeman, Oxford.

Rawlinson, F. and Farrell, P. (2008) Construction: a culture for concern? In A.R.J. Dainty (ed.), *Proceedings of the 24th Annual ARCOM Conference*, pp. 1093–102. Association of Researchers in Construction Management, Cardiff.

Red Tape Challenge (2015) Homepage [Online]. Available: http://www.redtapechallenge.cabinetoffice.gov.uk/home/index/ [25 September 2015].

Ridley, J. and Channing, J. (2008) *Safety at Work*. 7th edn. Butterworth Heinemann, Oxford.

Saari, J. (1994) When does behaviour modification prevent accidents? *Leadership and Organization Development Journal*, **15**(5), 11–15.

Skinner, B.F. (1978) *Reflections on Behaviourism and Society*. Prentice-Hall, New Jersey.

Starr, C. (1969) Social benefit versus technological risk. *Science*, **165**(3899), 1232–8.

Strauch, B. (2004) *Investigating Human Error: Incidents, Accidents and Complex Systems*. Ashgate, Aldershot.

Townsend, A.S. (2013) *Safety Can't Be Measured: An Evidence-based Approach to Improving Risk Reduction*. Gower, Farnham.

Weatherell, M. and Potter, J. (1992) *Mapping the Language of Racism – Discourse and the Legitimation of Exploitation*. Harvester Wheatsheaf, Hertfordshire.

Williams, K. (2006) State of fear: Britain's 'compensation culture' reviewed. *Legal Studies*, **25**(3), 499–514.

Wiggins, S. and Potter, J. (2007) Discursive psychology. In C. Willig and W. Stainton-Rogers (eds), *The Sage Handbook of Qualitative Research in Psychology*. Sage, London.

Woodcock A., Bently, D. and Glaze, B. (2012) David Cameron: I will kill off safety culture. *Independent*, 5th January.

Young, D.I. (2010) *Common Sense Common Safety*. Cabinet Office, HM Government, London.

Chapter Four
Safety in Construction

There have been significant reductions in the numbers and rates of injury over the last 20 years or more. Nevertheless, construction remains a high risk industry.
                    Health and Safety Executive 2015a

People get hurt and die on construction sites in the UK, and indeed all over the world. I have worked on sites where people haven't gone home that day – they have ended up in a hospital bed instead, and we've had a collection to help tide the family over whilst recovery takes place, as no wages will be coming in for the next few months. Safety matters. And it *really* matters to individuals and their families.

But we know this. In the UK, large construction companies, local agencies and the government invest heavily in terms of time and money to seek to improve safety on UK construction sites. Consequently, management of safety on large sites has developed complex approaches to measurement and analysis, and planning and control, to focus directions for positive change. Organisations are making significant efforts through the use of safety management systems, safety training and measures of safety competence to improve safety on their projects.

However, such approaches have to fit in with existing shared ideas of safety on sites, drawing on what safety already *is* to help

Unpacking Construction Site Safety, First Edition. Dr Fred Sherratt.
© 2016 John & Wiley Sons, Ltd. Published 2016 by John & Wiley Sons, Ltd.

develop and perpetuate any new thinking. When the usual ways in which we analyse and manage safety in construction are unpacked, our shared understandings around safety are revealed, which will themselves influence how safety works on such sites.

Measuring Safety: Accidents and Statistics

Nearly every book about construction safety starts with the statistics. Accident statistics are often one of the most common ways we talk about safety, and they have already appeared in this book in Chapter 2. In the UK, the Health and Safety Executive provides a complete annual statistical evaluation of the more hazardous industries, of which construction is one, readily available on its website. Often termed 'lagging indicators', accident statistics have long been one of the most readily available and understandable 'measures' of safety, and are frequently used to make broad assessments of an industry or project's performance.

However, the regular and repeated use of statistical data inevitably brings with it some misnomers about safety. For example, as Laitinen *et al.* (1999) argue, although often held up as such, a site with no accidents does not prove itself 'safer' than a site that has had a several accidents; random variation plays a significant part, and it also depends on where you draw the line – whether monthly, project or organisational data are used, and whether major incidents, minor incidents or near misses are included. As a result other 'leading indicators', such as levels of training or even safety climate questionnaires, are often used to provide a more holistic 'measure' of safety (Hinze *et al.* 2013). But no matter what other factors are used to calculate 'safety performance', the statistics remain most prominent; the number of hours worked without a reportable accident is often proudly displayed on the project hoardings of large sites, as shown in Figure 4.1, as well as the display of collated statistics on organisational websites.

Yet despite this being a clear demonstration of a *lack* of accidents, such an approach that grounds safety in numbers inherently contributes to the shared understanding that accidents *are* likely to happen on sites – that they haven't here (yet) is something to promote and use as advertising. Although this is of course something to celebrate, it does not help to shift thinking away from a certain level of inevitability that is embedded alongside our ideas of accidents, and how they fit within construction site life.

Figure 4.1 Our Safety Performance

Understandings of occupational accidents first emerged when the industrial revolution began to bring significant change to work and work practices at the turn of the nineteenth century. Before then, safety did not form part of the social context – accidents were seen as the simple consequence of an 'act of God' or attributable to individual negligence (Cooter 1997). Ideas of safety did change with the growth in industry and the development of mechanised factories, but the relationship was not one of simple cause and effect. Rather accidents, or ideas of unsafety, became normalised and legitimised – considered part of the 'natural order of industrial society' (Cooter and Luckin 1997: 5). Safety as part of industrial work processes was segregated from their development, seen as secondary to industrial and economic progress, rather than something that should be considered an inherent part of the development of new machinery and systems of operation themselves. As Cooter and Luckin (1997) go on to argue, the very act of labelling occupational injuries and fatalities 'accidents', shifts blame away from the labour processes and work, and puts them back into the realms of God, fate and individual responsibility.

In fact, back in 1844 when arguments around safety legislation for factories were ongoing in the House of Commons, construction was not even considered as an industry that *could* be improved with regard to safety, unlike work in factories or mills. Rather, people were simply expected to die on construction projects, and if they became a mason or joiner they would readily accept that risk as part of their trade (Warburton 1844). Accident statistics as a proportion of the workforce emerged as a justifiable and relevant consideration when assessing the human impacts of an industry – and they are still calculated and reported in this same way by the Health and Safety Executive today in its annual statistical summaries.

Accident statistics are also highly impersonal; aggregating accidents removes the people from the numbers – the workers become lost in the data. As McEvoy (1997: 75) suggests, 'they become an abstract factor of production, to be bought at a price determined by impersonal market forces and combined with capital and resources at the will of the entrepreneur'. But when people speak about safety, although they often still talk of accidents, they do so in a different way to this bald numerical approach. For the construction workforce, accidents are often a highly personal event – it is sadly very likely that someone with many years in the industry will have been witness to an accident, a near miss, or even been party to one themselves. The way people easily talk about such events illustrates how accidents are an inherent part of construction site life, for example:

> I've had a couple of very close escapes, nearly lost my foot a few weeks back.

A subcontractor's foreman casually inserted this potentially horrific incident quite neutrally within a conversation we were having about safety. There was no emphasis or expanded discussion about the incident. In fact, no details at all were forthcoming in this conversation, beyond the specification of the precise body part he was almost relieved of. We both simply accepted it, sharing ideas of acceptance and resignation – this was just part and parcel of working in construction. I could have told a few similar tales myself. But that the construction industry has got to this point is really very disheartening. Despite the potential to dismiss such comments as macho posturing, or even to add comic value to a conversation, that accidents are *so* commonplace that they are no longer unusual events and can be readily employed in such a casual way suggests at least some level of tolerance. In fact, in all the many conversations I've had around safety, accidents quickly appear as illustrative of safety, or rather unsafety, on sites. Such 'war stories' form a part of site life. The sad thing is that they are only told by those who survived to tell the tale. Ideas of inevitability and acceptance have developed over time into a shared resignation that accidents are simply part and parcel of construction work and construction site life. And the accident statistics of our industry do little to challenge this; despite improvements, accidents *do* happen – and we are only too aware of these numbers.

Although accidents are just one of the ways in which we construct safety on sites, they help create a shared acceptance of a reality in which accidents *will* happen. It is in this reality where safety improvements are sought. But this is not a context which readily supports a shift to a safer industry. Indeed, an inherent fatalism has often been identified within industrial workforces which can play havoc with organisational safety targets, particularly those around zero – currently the biggest number in construction site safety – and one which is explored in much more detail in Chapter 8.

The more we reiterate accidents and statistics, even in the negative formats of 'x hours worked without ...', the more we support the understanding that this *is* inevitably a dangerous industry, and one in which accidents are just a familiar characteristic of the construction site environment. We could instead do worse than recognise the roots of 'occupational accidents' as a direct result of our work process and practices, rather than inevitable consequences, and therefore start to challenge the *way* we do things from new perspectives. That we so readily talk about accidents and not safety also helps us understand safety a little better – it is paradoxical, much more easily identified and understood by its absence than its presence, and accidents are the ultimate manifestation of unsafety in practice.

Cause and Effect

Another characteristic of accidents is the way they are often explored through ideas of cause and effect – put most simply unsafety is the cause, and the accident is the effect. Looking back over historical safety statistics, it is not surprising that recent improvements in safety on construction sites have been recognised:

> A few years ago I noticed a lot more accidents, people breaking limbs and falling off things ... there's a lot less accidents now ...

As this foreman notes, there has been a change – things are getting better, although accidents still do happen, there are less of them, something he ascribes to:

The attention being made to safety.

Here, accidents are being used as 'evidence' of safety, both in the past and present, and safety is the reason for the reduction. This creates another cause and effect scenario; improved attention to safety resulting in fewer accidents.

These ideas of causality also imply that accidents can be used for learning and improvements around safety. At their simplest level, statistics provide focus for interventions; that 45% of workers killed in the construction industry during 2013/14 fell from height (Health and Safety Executive 2014) justifiably suggests this is still an area worthy of directed management efforts.

More complexly, accident investigation becomes part of the process for understanding why accidents occur on sites and therefore how future performance can be improved (for example see research by Ahmad and Gibb 2004; Chua and Goh 2004; Manu et al. 2010, although as Townsend (2013: 117) notes, models of accident causation do seem to continuously change and develop). Hollnagel (2004) suggests this also satisfies desires for certainty in our worlds, and the idea that knowledge has been gained which can be used in future accident prevention. In some approaches this has evolved into a quest for root causes, although at times this is in danger of becoming a thinly disguised quest for blame (Whittingham 2004). As Dekker (2011) has argued, if accidents are seen as evidence of error or failure, accident investigation becomes the quest to identify the responsible individual behind the error. This can be seen in reports of the Crossrail fatality of 2014, the Chairman stating that:

This individual went into an area that was excluded, shouldn't have been in there and we're still trying to understand why it happened (as reported in the Independent 2015).

Here, although understanding has been positioned as the ultimate output of the accident investigation process, it is the worker who is effectively 'blamed' for the incident, his action becoming the focus of the investigation process. There is a segregation of responsibility for this incident; the worker's violation is separated and made distinct from the shared ownership of the accident as a whole.

Statements like this are common, and continue to perpetuate the ideas of human error as a prominent 'causal factor' in accidents. Linking back to cognitive ways of thinking about people, the 'cause' of the error can become easily identifiable as one of Reason's (1990) famous rule, skill and knowledge-based errors, or occasional or routine violations – the Crossrail fatality could potentially have been ascribed to any one of these. Yet this is also perpetuating simplistic ways of thinking about accidents, seeing them as a straightforward sequential 'casual chain' into which human error has a clearly definable place.

Causal thinking around accidents has a long history, Heinrich's 'accident pyramid' has made an appearance in almost every large construction contractor's boardroom during a health and safety presentation, a variety of numbers allocated to the near misses at the base, up through minors and majors to one fatality at the top. Yet as Hollnagel (2014: 71) wryly states, beware the 'allure of graphical representation'. Heinrich's pyramid has been criticised for its suggestions of direct causality and its implied relationships, and how it is commonly used in safety training, despite Heinrich's own caveats and warnings about using his data in this way. Townsend (2013: XIV) goes further to suggest that 'the original research upon which much of today's safety management is based is at best fragmentary; at worst it is spurious'. Yet such issues aside, there remains an ongoing quest to report as many near misses as possible on sites, as if this will somehow prevent accidents occurring higher up the pyramid. Setting quotas for near-miss reporting has become a key performance indicator (KPI) of its own, despite the obvious temptation for those responsible for delivery of these quotas to actively seek out (and even produce) any shortfall.

The continued faith in near-miss reporting ignores the fact there have been significant advances in accident thinking that challenge these simplistic cause and effect approaches. For example within his *Field Guide to Understanding Human Error*, Dekker (2006) argues that accidents are only symptoms of deeper organisational causes, such as time and money pressures leading to unsafe acts on the part of the workforce; mistakes forced by these 'latent defects' in work structuring and management rather than the simplistic excuse of errors made by the individuals concerned. Hollnagel (2014: 63) actually presents the idea of causality as a 'Safety-I myth', arguing that the idea of backwards causality from effect to cause is logically invalid; if a cause does exist, it does not necessarily mean it can or will be found. He also argues that causal thinking relies on the cognitive notion of the 'rationality assumption', although people are *irrational*, and furthermore would have to operate in 'a deterministic world that does not really exist' (Hollnagel 2014: 64). Although he has not quite said it, Hollnagel's challenge is grounded in social constructionist ways of thinking about people; logic is not an option and causality is not really a viable tool for analysis in what is a messy and incoherent world. Dekker and Hollnagel explore these ideas in much more detail in their own books, which are recommended for anyone wishing to learn more about these different ways of thinking about safety, rather than to do these authors the disservice of condensing them into a poor repetition here.

Yet despite such challenges, historic and familiar ways of thinking about accidents, such as the accident pyramid, have helped socially construct safety through ideas of causality, developed by the inclusion and role of people within the resultant causal chains. As suggested by the Chairman of Crossrail in his public statement; a cause can be identified, blame apportioned, and lessons (should be) learnt. As a result, measurement, causality, learning and improved understanding as facets of accident statistics and accident investigation remain prominent within wider understandings of safety.

However, such approaches may simply be misdirecting our efforts towards the allocation of blame and the production of reams of near-miss paperwork, rather than taking a closer look at how we structure and manage our work contractually – to look beyond over-simplistic causal chains and instead focus more closely on the bigger picture – the construction site contexts of Chapter 2. We need to ensure that the 'hidden' influences of time and money, the manifestation of Dekker's (2006) latent defects in our construction industry systems of work, are put in their correct place within the cause and effect chains, and in the way we think about safety. We need to shift our understandings away from the 'sharp end' at the site level, and move them up into the boardrooms where decisions to take a slice off the tender price or a month or two off the programme can result in pressures for speed and productivity that creates an unhealthy context for work on sites. The way this thinking manifests in site practice, examined from the perspectives of the workers themselves, is explored in much more detail in Chapter 6.

Safety Management Systems

Most large construction sites will be operating a safety management system. These provide structured guidance on management actions and procedures, templates for inductions and inspections, as well as various operational systems to ensure all legislative requirements are met. However, by its very creation, the idea of a safety management system fundamentally positions safety as something that now has, and indeed *can* have, its own management system and can therefore actually *be* managed on sites. However, this way of thinking ignores the fact that it is not actually *safety* that needs management – but unsafety. Drawing on the ideas of Weick, Hollnagel (2014: 5) suggests that safety is itself actually a 'dynamic non-event'; it is the state of something *not* happening. Therefore a safety management system is essentially a misdirection – and in looking to manage 'safety' we are in danger of focusing on something that is not actually even there.

But as Hollnagel (2014: 6) then goes on to state, this definition is 'very clever ... but it introduces the small problem of how to count or even notice or detect a non-event', and this is where safety becomes challenged within safety management systems – how to manage something that is not actually happening?

Well, there are certainly no shortages of suggestions of how to do this, and what a safety management system should contain in order to achieve it. Many textbooks and consultancies can provide checklists and pro-formas to create a coherent safety management system, yet such a formulaic approach can be seen as a little incoherent for the management of something that doesn't actually happen, within the context of a messy, inconsistent, and constantly changing social reality.

Early guidance from the Health and Safety Executive around safety management systems is of interest here, precisely because of how it has changed in line with developments around safety management thinking, some of which resonates with the social constructionist ideas of safety explored in this book. The 2006 edition of HSG65 was titled *Successful Health and Safety Management* (Health and Safety Executive 2006), drawing on the idea of safety as something which can be managed, and sought to prescribe its 'success' through a mechanistic approach. However, the revised 2013 edition suggests that such an approach has proved problematic and ineffective; instead the Health and Safety Executive (2013:7) articulate the need to:

> ... treat health and safety management as an integral part of good management generally, rather than a stand-alone system.

Although safety remains (perhaps unsurprisingly) bundled up with health, here it is actively positioned as a part of management as a whole. Furthermore, the title of the revised HSG65, *Managing for Health and Safety*, also makes a significant change in how safety is positioned within the management context. Whilst the 2006 edition suggested that safety *could* be managed, and indeed that there would be clear indications and the ability to achieve success, despite the problems of measuring a non-event, the 2013 edition has subtly shifted in its own understandings. Management is now something that contributes *to* safety; safety becoming the *outcome* rather than an entity to be managed itself. This change in the action and positioning of management within this relationship is highly significant, and develops safety as something that is produced rather than something that has action (or rather management) simply done to it.

Although a relatively minor aspect of the much more significant changes found within the revised 2013 edition of HSG65, the way safety is positioned within the title of the document is itself important. It will necessarily contribute to the way safety is understood through the sharing of what is arguably one of the key guidance documents addressing safety in the UK. Yet, although such changes have been made by the Health and Safety Executive, given the well-established ideas and ways of understanding safety and the principles of safety management, and indeed the reliance on some form of safety management system, it may be some time before such integration and the repositioning of safety can be achieved. Indeed, the fact that the HSG65 still exists in part retains and perpetuates the understandings of safety as a separate entity, born of the inevitability that segregating safety out for prioritisation will always disassociate it from practice and any wider work contexts.

This has inevitably led to various challenges and conflicts between safety as it can be managed theoretically through a safety management system, and how safety works in practice; for example around the complexities of change within the site environment and how best to position the 'management' of such a changing space (see Chapter 5), the need for rules and their enforcement alongside worker engagement (see Chapter 7) and the ever-attractive approaches of 'measurement' (see Chapter 8). Although safety management systems have been successfully implemented by many large construction contractors (Health and Safety Executive 2009a), such challenges may well have been realised. The Donaghy Report (2009), *One Death is too Many*, concluded that although safety management systems were often found to be strong within the corporate core of the organisation, management to the very ends of the supply chains and implementation on the sites themselves was not always as successful (Health and Safety Executive 2009b). Despite best efforts, the shared ideas of safety within the safety management system are not necessarily those found on sites, where the actual management and work occurs.

A final, yet fundamental point to be made with regard to safety management systems is that their all-encompassing nature often means that safety can easily become cluttered with inspections, rules, objectives and methods for measurement. Yet when we look beyond their titles, within the Health and Safety Executive's safety management guidance, swift prioritisation is made of the ideas of hazard, harm and risk; something that often forms only a *part* of safety management systems.

Indeed within the Health and Safety Executive (2015b: 7) guidance for the Construction (Design and Management) Regulations 2015, the first of the 'key elements to securing construction health and safety' is:

Managing the risk by applying the general principles of prevention.

Within safety management systems, the focus on safety means that hazards and risks, the elements that actually challenge the non-event of safety, are sometimes neglected. Implementation of a safety management system does not inherently cascade hazard and risk along with it, rather 'safety' is prioritised, and when placed alongside 'management', it can be further segregated from construction work itself. The management of safety through a safety management system therefore becomes a much more limited construct, and one that can be both reinterpreted and misinterpreted in practice; it would arguably be far more appropriate to create a Hazard Management System or Risk Management System, focused on the management of unsafety; if we look after the hazards and risks, then safety will take care of itself.

Competence

Manager, supervisor and worker competence and training are all, to various extents, required by legislation, promoted by government initiatives and incorporated within safety management systems. It is argued that to enable operatives to work safely, they need to be trained and equipped with the skills to make them competent to carry out their tasks (Teo *et al.* 2005). However, despite this fundamental acceptance, and repeated references to competence and training within regulations and guidance, it has been suggested that there is no clear standard or benchmark for what people *should* know about safety on sites (Health and Safety Executive 2009c) – in fact, what *is* competence, and how does it work with safety?

Shared understandings and implicit 'definitions' have emerged around competence, which have made any clear and agreed idea something of a challenge for the UK construction industry. Historically apprenticeships and on-the-job learning over a period of time enabled those entering the industry to obtain and hone their construction trade skills and eventually emerge as competent workers – 'time served' considered a validating process, despite the obvious potential for it to be no such thing. Such thinking can also be seen in the ideas of competence as an amalgamation of such traits, for example the mnemonic KATE, which brings together Knowledge, Ability (or Attitude or Aptitude in some versions), Training and Experience as a 'measure' of whether someone is competent or not. But despite such attempts to qualify competence and position it clearly within industrial parameters, it still remains a rather intangible human characteristic.

Safety is also considered through ideas of competence. For example research by Cipolla *et al.* (2006) in Australia established that a certain level of 'competence' is required within the supervisory structure, to ensure that people in 'safety critical' positions have the knowledge and understanding to develop the correct workplace environment. The Health and Safety Executive (2009c) found that a lack of competence can result in a lack of confidence, and result in either excessive safety management or disinterest in safety management amongst supervisors and managers. But despite the fact that 'competence' has become a central concept to safety, and common phraseology within safety literature and guidance, it has never really been clearly defined.

In fact, evidence of the problems around understandings of competence as a measure of safety can be seen in the way the UK has changed its safety legislation. Although the Construction (Design and Management) Regulations of 2007 directly referred to competence in Clause 4:

> ... unless the person to be appointed or engaged is competent

no clarification of what competence is was included within Clause (2) Interpretation, precisely where definitions of all the more subjective terms within the legislation itself should be found. In order to support this, the Health and Safety Executive's Approved Code of Practice for the 2007 Regulations tried to clarify competence within the professional team as:

> (a) sufficient knowledge of the specific task to be undertaken and the risks which the work will entail;

> (b) sufficient experience and ability to carry out their duties in relation to the project; to recognise their limitations and take appropriate action in order to prevent harm to those carrying out construction work, or those affected by the work (Health and Safety Executive 2007: 45).

Thereby making recourse to further sub-definitions and categorisations, as found under KATE. Within the guidance, this 'definition' was further supported by recommendations for the assessment of competence within the workforce itself. However, these recommendations were themselves limited to establishing operatives have a basic knowledge and understanding of tasks, and that this knowledge is regularly updated through training via tool-box talks or more formal training programmes such as National Vocational Qualifications (NVQs). It has also been identified that there is a general understanding that competence relates more to knowledge and experience, rather than such specific skills or qualifications (Hughes and Ferrett 2007). These problems with creating an accepted 'definition' for use in legislation can only hint at the complexities that will underlie and contribute to any shared social understandings of safety competence out on sites. Indeed, research of competence and training carried out by Ness (2010) from a constructionist perspective found that personal acquaintance, recommendations from other workers or an informal trial period of work were actually far more relied upon as measures of competence than any form of qualification or training. Indeed, Ness's work also highlighted the importance given to the inspection of a construction worker's tools as a measure of their competence – new tools suggesting a lack of experience, a 'chancer' rather than a 'proper' construction worker, whilst worn yet quality tools that have been well-cared for prove a sure measure of competence and skill. Yet how to include this informal, but probably highly accurate 'measure' of competence within a mechanistic safety management system checklist would be something of a challenge.

The problematic nature of competence is also reflected in the changes made in the Construction (Design and Management) Regulations for 2015, which have actually removed any reference to competence within this revised edition. Clause 4 (as was) has been omitted and competent still does not feature in Clause 2 Interpretation. However, within the Health and Safety Executive Guidance (2015b: 8), those appointing the professional team must ensure they:

> Have the skills, knowledge and experience to carry out the work in a way that secures health and safety.

Competence has now been exchanged for more detailed components: skills, knowledge and experience, although none of these are themselves defined further in the Guidance. Safety here is again something to be secured, positioned as the outcome of work processes, rather than a separate entity to be managed apart from the business of construction.

For those appointing anyone to work on a construction site, a slightly different 'version' of competence is included, and the workforce must have:

> The right skills, knowledge, training and experience to carry out the work ... in a way that secures health and safety for anyone working on the site.

Important here are the 'right' qualities, which necessarily creates the acknowledgement that there are those that may be 'wrong'. In associating 'right' with the entity of safety, wrong becomes suggestive of unsafety, and this revised guidance now positions skills, knowledge, training and experience as more fluid, along a right/wrong continuum. No matter that someone has 25 years' experience bricklaying; it does not mean they are able to bricklay safely on site. The 'right skills' are perhaps more reflective of more intangible measures such as the quality and state of a worker's tools, considered alongside the way they use them.

Training has also been included here (where it is not for the professional team), and has now been elevated from a tick-box exercise of achievement, to something more subjective that should also be evaluated in terms of its relevance to safety in practice. Yet, at the time of writing, these changes have only been made very recently within the safety management legislation of the UK, and their impact and influence on the way we think about safety have yet to be realised. However, the potential for change is clearly emerging, and this revised legislative approach enables challenges to be made of existing ideas of 'competence' with relevance to construction work, and more considered reflections of how 'safety competence' may actually work in practice.

Training

Safety training has often been considered vital in the quest to improve safety on sites. There are two differing approaches to site safety training: classroom training and on-the-job training, with different underlying aims. Formal, classroom training has been suggested to seek to improve individuals' awareness, knowledge and understanding of safety on sites, and research by Hare and Cameron (2010) suggested that such safety training for managers is an effective method of improving competence, even linked to lower accident rates. On-the-job training is more often seen as attempting to achieve a positive behavioural change, by changing attitudes in the real-life context. Yet both methods have received criticism: training is not necessarily successful in achieving its aims without validation and active evaluation (Lingard 2004); and doubt has been raised as to the extent of influence classroom-based training actually has on work in the site environment (Kamardeen 2011).

Informal training methods have also frequently been found to operate on construction sites – not least within apprenticeships. Despite formal training, Rooke and Clarke (2005) found that many operatives feel they learnt more on the job by watching more experienced workers, trying things out or by direct instruction. This trial and error approach by both new and experienced operatives will result in the sharing of understandings around safety and the perpetuation of existing practice.

As one operative said:

When you do an IPAF or PASMA at least you've getting your hands dirty, not just a sit down in front of a video.

The training referred to above for both mechanical and non-mechanical access equipment is necessary for specific activities on site, and results in the acquisition of 'tickets' – the site term for training certificates. Here, the practice of acquiring these tickets has been associated with a direct interaction with the site environment; 'getting your hands dirty'. This training is an active process, positioned against the passive 'sit down' approach of classroom-based training. This suggests a clear distinction between training as associated with practice and finding out how to do things, and education, carried out away from the active site context. A value judgement has also been placed on the process, the practical considered more beneficial than the purely 'academic', suggesting the way any impractical safety training is likely to be accepted and received.

Research carried out on the London 2012 Olympic Park also found that safety training for the workforce was most successful through verbal, face-to-face talks such as daily briefings, inductions and tool-box talks (Hartley *et al.* 2011). Site inductions are a legal requirement under CDM2015 Clause 13(4), and are required for all new personnel before they commence work on site. The aim of the induction is to impart and educate new operatives of the safety requirements, and so should include basic site information such as the location of welfare facilities and accident reporting procedures alongside key risks and controls such as permits to work, traffic routes and hearing protection zones. Tool-box talks are also common within larger organisations and are often delivered weekly around a specific topic, relevant to workers' current or forthcoming tasks, such as safe systems of work or use of specialist personal protective equipment (PPE). There have also been developments with the use of pictorial-only systems for non-English-speaking operatives, and these have actually been found to be more effective than text-only safety training in tool-box talks for all workers (Hare and Cameron 2011). Often these approaches are prescribed by the safety management system, which can dictate format, content and delivery mechanisms.

However, whilst such communications may be effective in theory, there can be challenges in practice. There is the potential for inductions to become bloated and repetitious, as sites aim to present all the safety information ever prepared by both the managing organisation and its project team, or to reiterate all relevant legislation in full. Arguably inductions should contain only what is necessary and relevant to a particular project, as the Health and Safety Executive (2009d) say in its guidance example:

> You have probably gone through hundreds of site inductions and will probably go through hundreds more.
> The induction is important as all sites are different and have a wide range of hazards which will change as the site develops.

As with risk assessments and paperwork, safety inductions are often in danger of becoming more associated with time – and more accurately the waste of it – than any more positive experience; thereby establishing a conflict between safety and work from the very first moment a worker steps onto site. Efforts are being made to develop inductions in a more relevant and appropriate format, instead of the stilted delivery of a set of slides by the youngest member of the management team, or presentation of an out-of-date video focused on groundworks and foundations when there are now three tower cranes and a 20-storey building on the site. Inductions contribute to the social construction of safety through training, and long and unnecessary sessions, delivered by the person who drew the short straw, do not prioritise safety as a necessary and integral part of work practice. Instead they all too often help in its dismissal as an inconvenient part of construction working life, simply reinforcing the derision of safety in society that has also been brought into the induction room rolled up inside the morning paper.

Tool-box talks also have the potential to associate safety with time wasting and irrelevance to practice, should a formalised safety management system timetable or checklist for delivery result in the delivery of 'Sun Safety' during a particularly wet October, or that 'Deep Excavations' be delivered to the floor layers on a project a week from handover. Again, this creates a shared understanding of safety as something irrelevant, a thief of time, and something that gets in the way of work to be done – explored further in Chapter 6.

More formal safety training can lead to the award of qualifications. Whilst many professional and trade qualifications vary in the extent to which they teach safety skills and understanding within their content, there are some specific safety-only qualifications that have become standard in the UK industry. The standard for site operatives is the contentious Construction Skills Certification Scheme (CSCS) card. The CSCS was recommended as a minimum requirement for all site operatives as evidence of 'competence' under the CDM 2007 Regulations, as a way to ensure a basic knowledge and understanding of safety on sites. CSCS cards are also linked to NVQs, which are themselves awarded for the demonstration of site-based competences through practical tasks and skills (CSCS 2015). The CSCS card is trade-specific, and issued according to an individual's work experience and training. There are also affiliated certification schemes which require a higher standard or qualification such as the certification for plant operators, the Construction Plant Certification Scheme (CPCS), which incorporates theory tests, on-site assessment and ongoing records of progression, or the scaffolders' Construction Industry Scaffolders Record Scheme (CISRS) card which can only be gained through an intensive training and the NVQ programme (Construction Industry Training Board 2015).

However, under the 2015 CDM revisions and the shift away from an overarching ideal of 'competence', the CSCS card is no longer championed as it once was. Indeed in the 2015 CDM Guidance the Health and Safety Executive (2015b) states that:

Sole reliance should not be placed on industry certification cards.

This is likely to be as a direct result of the criticisms CSCS has attracted over the years. It was seen by some within the industry and the unions as creating a carded, rather than competent, workforce (Spanswick 2007), and there has been little published evidence to link the CSCS card to overall site safety performance.

Nevertheless, CSCS cards are now the industry standard and an essential requirement for access to work on major contractors' sites (CSCS 2015). For example they were a mandatory requirement to work on the Olympic Park site (Richardson 2006). Many inductions will state that:

You should have your CSCS/CPCS card ready for inspection.

Making the assumption that the audience hold and are currently in possession of their CSCS cards and tickets when they arrive on site. This positions CSCS cards as a central part of safety training, and by association, with safety competence within the site environment.

Yet this does not consider the ease of obtaining a CSCS card, legitimately or otherwise.

However, it must be recognised that such training extends well beyond safety, despite the often prominent inclusion of safety within the training programmes. Often safety is not realised or even associated with training or tickets, instead as one operative suggested:

> You've got to keep up training, if people's tickets have run out the company need to keep them up for the workers.

Within this conversation, training was originally being positioned by the speaker as a way of ensuring safety on site – the ideas of training integral to his understandings of safety. However, talk of training then shifted away from safety itself to a much more straightforward association with practice. Here, training is undertaken for the benefit of the operatives, but not positioned alongside safe practice or the knowledge gained during the training process, rather it is positioned amongst the consequences when tickets 'run out'. This operative is more concerned about operatives whose tickets have expired, which would in practice limit their involvement in certain activities of the site, and therefore their potential to earn money as a qualified worker, rather than any direct association to safety. In contrast to the intentions of many safety management systems, which seek to embed safety within all aspects of practice and associated interactions on sites, site understandings of training are more often focused on certain tasks and activities. Training becomes associated with initial access to the site, specific practices and the practicalities of site activities. The consequences of possession of such tickets is often emphasised with reference to how many machines they can drive or tools they can operate and so what money they can earn, rather than the increased safety of those activities subsequent to training.

Alternatively, this could actually be the manifestation of the ideal 'version' of safety on sites – it has become something completely inherent in construction work, and so does not need to be separated out from the business of operating machinery or tools. As CDM 2015 desires, safety is now embedded in the *right* skills and training for the work. However, whilst this would be an ideal scenario, the fact that practice dominates the discussion, along with *payment* for this practice, suggests further exploration of this relationship between safety and work is needed, and this is duly examined in Chapter 6.

Personal Protective Equipment

One of the most common ways people understand safety is through personal protective equipment (PPE). PPE is so dominant a feature of our industry that it even makes a key contribution to the social construction of the construction worker – the hard hat and yellow vest assigning identity and associated status, easily recognisable by the rest of the world. On sites, shared understandings between the workforce enable PPE to communicate even more; harnesses for scaffolders and roofers, orange vests for banksmen, blue or black hard hats for foremen and supervisors. It is also by its very nature highly visible; PPE is *meant* to be seen.

However, this has led to a substitution of PPE for safety itself, a consequence at direct odds with accepted risk assessment philosophy. Whichever 'hierarchy of risk reduction' is utilised, PPE is always near the bottom in terms of a risk reduction strategy, if not actually in last place. The Health and Safety Executive (2003) states that PPE should be the 'last resort' in health and safety management, yet its visibility and prominence within the construction site environment has led to a social construction of safety so tightly interwoven with PPE that the two are rarely separated. Although PPE is certainly an *artefact* of safety, it is *not* safety itself, and it is important that this distinction should be explored to ensure it is better understood.

PPE first appears at the gates of our sites; through both text and iconography, minimum PPE requirements are emblazoned on signage with the function of ensuring compliance with basic site safety requirements, through the stimulation of immediate action from their audience. Signs often ask that:

> Beyond this point the following PPE MUST be worn as a minimum …

Such signs demand compliance with the site safety requirements of PPE, establishing the minimum standards required for all those working on the project, but in doing so they also set up several other considerations and shared ideas about PPE and its use in practice.

For example, the sign defines a distinct space by its presence – it is 'beyond this point' that the PPE listed must be worn. Compliance with the sign is bound up with the actions of its audience to wear PPE in order to pass this 'boundary', however, rarely is any allowance made for non-compliance and no punishments are usually identified. It also creates a 'practice' of PPE, PPE must be 'worn', physicality is involved, and furthermore it acknowledges that there may be more PPE to come, the items illustrated on the sign are simply a minimum. Such texts are often accompanied by images – items of PPE contained within the blue circles of mandatory compliance iconography.

However, such signs also assume various shared understandings within their audiences by what they do not say. Here, there is no spelling out or explanation of PPE, its meaning and understanding is assumed, as evidenced by the items that follow. Although PPE is something inherent in practice, it can be surprising how many people, who would consider themselves to be working in the construction industry, do not fully understand its meaning – PPE seems to be something you have to have worn a few times to fully appreciate. Therefore this lack of articulation does not help those who are new to construction or even just new to working on a large site, rather than small projects where trainers and a t-shirt will do just fine.

Why this is only the *minimum* PPE is not explained, and the sign here is assuming understandings of risk assessments and task-specific PPE without further articulation. Yet to make such assumptions is arguably bad management practice – and again for those new to the project or the industry, this does not position PPE as a valuable tool of safety management. Rather, PPE has simply become a set of objects, something vaguely relating to safety, which must be worn in order to pass a boundary – why you need to wear them, or what happens should you decide to keep them on or not remains a mystery.

Different signs position PPE in different ways, linguistic ambiguity potentially at odds with the overall demands of the use of PPE in practice, and the potential variations in the understandings of the audience. For example, PPE may need to be worn, or people can simply be 'equipped', it can be a defining factor of practice – the familiar off-the-shelf sign:

No PPE No Work!

Often this is combined as above with notions of demarcation and boundary. Articulation of the disciplinary process is often simply summarised as 'No Work!' but in the case of non-compliance once work has commenced, any repercussions are often missing.

Indeed, formal signage around PPE often lacks any provision for non-compliance in terms of discipline or punishment. Instead, they assume to operate within a reality where everyone complies with PPE requirements. However, this means that any need for discipline and compliance cascades down the site management team; the more mundane aspects of the management of PPE in practice are found on 'site-made' signs – handwritten or printed and laminated in the office. Such signs often clearly articulate disciplinary repercussions for non-compliance, threats of re-induction or removal from site can be boldly stated. In such contexts PPE is often reduced to one of 'the rules' – rather than a positive and proactive method of safety management.

In keeping with more formal signage found on site entrances, the *actual* role PPE takes in the support of safety on sites is assumed to be fundamental shared knowledge within the workforce, despite the fact that this may not be the case. This in part explains the ongoing problems that many site managers and foremen have with ensuring PPE compliance on sites; if PPE is not understood, and not accepted by the workforce, the current way it is positioned as part of construction site safety is not likely to change such attitudes or facilitate compliance. Such issues of enforcement and engagement, and the different roles the site hierarchies play in safety management is explored in more detail in Chapter 7.

The role of site management in the enforcement of basic PPE can also be seen in the ease with which it is used when we are talking about safety on sites. PPE is often the easiest thing to grasp, as this supervisor does when talking about unsafety:

> If someone's not got their glasses on, I'll tell them to put them on.

Here PPE, specifically light eye protection glasses and more specifically failure to wear them, has become the manifestation of safety management in practice. PPE becomes the most common way of talking about safety – possibly because it is simply so easy to do so. Rather than have to explain a complex situation involving work methods or situational hazards, PPE is a readily understood way of talking about safety – and is unsurprisingly therefore a very common aspect of daily safety management practice.

In fact, recent developments around PPE – notably the inclusion of light eye protection (safety glasses) and gloves as mandatory on many large projects – will have likely influenced the positioning of glasses as the most prominent contemporary PPE artefact of safety, as hard hats once were in the 1980s. Indeed, glasses as mandatory PPE have often been challenged in their practical use on construction sites. Researcher Martin Löwstedt (2014) even titled his participatory work on site 'Taking off my glasses in order to see', finding that they misted up as soon as he started to carry out any physical work, creating a more serious safety hazard than they could ever realistically hope to prevent. Although company directors, managers and health and safety inspectors are able to walk around site wearing glasses without problems, for those actually breaking a sweat misting up is a constant problem and may well explain the prominence of glasses as the most violated item of PPE:

> People will work safely, not that they won't take the odd risk, with glasses and things like that.

For this supervisor, PPE is again the easiest thing to draw upon when talking about safety violations, and again glasses are selected as the artefact of choice. Violations around PPE are here constructed as only a minor infringement, the 'odd risk' within the overall behaviours of 'work safely'. This minor status associated with PPE as safety is another common characteristic; PPE is not valued highly within the practice of safety despite its participation as the very manifestation of safety in practice from the site gates onwards.

Potentially, it is the mandatory nature of PPE that has resulted in its devaluation, and the need for site-level enforcement for compliance. The ever increasing list of PPE items to be compulsorily worn on sites removes another aspect of reflection and conscious consideration around safety, alongside judgement and autonomy, from those actually carrying out the work and who are themselves the ones at risk. That the communications of PPE often neglect to explain the reasons for their use and assume understandings within the workforce also contribute to this shared construction. PPE signage often instructs without informing, even when to do so would potentially increase compliance.

An example of this can be found in the way PPE is used within signage on sites, the mandatory blue circles showing what additional PPE is required within specific areas or for certain work practices, such as that found in Figure 4.2.

But in Figure 4.2, the sign has again used PPE with assumptions of shared understandings; although ear protectors must be worn, it does not explain why, or if a certain level of protection is needed. Will ear plugs do or is something more robust needed? I'm only walking through, do I need protection? Where is the boundary for this requirement – is it just nearby or for the whole floor plate? With noise you could well suggest that you should be able to hear it to answer these questions, but people don't have built-in decibel monitors, and this sign arguably does not help the workforce determine the levels of PPE needed, over what durations, and why. Here a safety hazard has been constructed through the PPE needed to mitigate it, but that is very restrictive. Whilst signage cannot explain everything – the writing would get too small for one thing – there is a danger that simply by communicating hazards

Figure 4.2 Ear Protectors Must be Worn in this Area

through PPE contributes to the idea that safety *is* PPE, rather than anything more.

Furthermore, signs such as that shown in Figure 4.2 that operate without making the workforce aware of the hazards arguably relieves them of any safety responsibility, and their participation in making decisions or evaluations about their own safety behaviours and the workplace; instead an anonymous authority demands somewhat incoherent action from the audience through the signage.

Safety as PPE is often the most convenient example, manifestation and crystallisation of safety within the common site repertoire. This could be due to the high visibility of the artefacts of PPE, which make it easily and quickly assessable in terms of safety compliance. No special knowledge or skills are required to ascertain whether people are wearing their basic PPE, which therefore makes it a straightforward assessment of safety in practice. This provides management with a simple 'measure of safety' which can be employed as a benchmark for enforcement, which may not be as personally or professionally threatening as, for example, being able to assess the 'safety' of a complex scaffold erection.

Yet, and arguably most importantly, this dominance of PPE within the talk of safety also serves to reinforce the (mis-)understanding that PPE actually *is* safety in practice. Concern for workers at height should focus on leading edges and restraint, rather than whether they have their glasses on or not, and this should also be the concern of the workers themselves. Education and explanation are often missing from the construction site environment, perhaps a legacy of more autocratic times. Although some managers risk upsetting their morning inductees by showing pictures of severed toes to emphasise the need for safety boots, in other cases PPE is simply enforced as if it *were* safety itself.

This suggests it is very necessary to rethink PPE; it should be firmly relegated to the last resort. Reflection is needed of the benefits of continuing to layer workers in 'mandatory' PPE that is not effective in practice, and which only serves to devalue understandings and adoption of necessary and essential PPE as a whole. There is nothing more likely to resonate with ideas of safety as stupid and illogical than a pair of safety glasses that you cannot actually see through. PPE will inevitably remain prominent in the constructions of safety – it is after all fluorescent for a reason – but it should be repositioned in its wider context, to ensure it is not devalued where it is necessary, nor used as a simplistic replacement for safety and safety management in practice.

Summary

The ways in which safety in construction is measured and managed are able to reveal insights into the nature of safety itself within this context. Although approaches to safety management seek to improve and enhance understandings of safety within the workforce on sites, in many cases they simply do not help create a receptive context for such change, and in some cases actually provide a direct challenge.

A focus on accidents, or more specifically their statistical aggregation, helps create and support ideas of inevitability, no matter that the numbers are used to 'promote' safety. Safety management systems by their very nature position safety as something to be managed, and although recent changes have sought to move safety into the role of outcome, the familiar safety management system title and practices are likely to still inform understandings, despite the need to actually manage unsafety in practice. Safety management systems have close links to training and education, and the highly complex notion of 'safety competence', something vulnerable to becoming a tickbox exercise through the acquisition of tickets and course certificates, more closely associated with earning capacity than safe work practices. Because of its high visibility, PPE often becomes safety itself, yet this can easily devalue and diminish its intended role in safety. Management become distracted in the enforcement of site PPE rules, which in turn relegates PPE to something more associated with compliance and punishment, rather than protection.

Yet it cannot be assumed that any particular 'safety' will cascade down the management and supply chains to the sites and then manifest, unchanged, as safety in practice. In fact, it is very unlikely that this is ever the case – the establishment of safety in the office does not necessitate its emergence on our sites. A better understanding of the complexities and incoherence of safety within the site environment is needed, and the rest of this book now seeks to explore safety on sites through the shared understandings of those who work on sites every day. Rather than measure policy or evaluate management systems, exploration is made of actual practice, and how people consider and position safety within their lived realities of the construction site environment.

References

Ahmad, K. and Gibb, A. (2004) Towards effective safety performance measurement: evaluation of existing techniques and proposals for the future. In S. Rowlinson (ed.), *Construction Safety Management Systems*, pp. 424–42. Spon Press, London.

Chua, D.K.H. and Goh, Y.M. (2004) Utilising the modified loss causation model for the codification and analysis of accident data. In S. Rowlinson (ed.), *Construction Safety Management Systems*, pp. 443–63. Spon Press, London.

Cipolla, D., Sheahan, V.L., Biggs, H. and Dingsdag, D. (2006) *Using Safety Culture to Overcome Market Force Influence on Construction Site Safety* [Online]. Available: http://eprints.qut.edu.au/3801/1/3801.pdf [20 September 2015].

Construction Industry Training Board (2015) *Construction Plant Competence Scheme (CPCS)*. [Online] Available: http://www.cskills.org/supportbusiness/cardschemes/availablecardschemes/cpcs.aspx [25 September 2015].

Cooter, R. (1997) The moment of the accident: culture, militarism and modernity in late-Victorian Britain. In R. Cooter and B. Luckin (eds), *Accidents in History: Injuries, Fatalities and Social Relations*, pp. 107–57. Editions Rodopi B.V., Amsterdam.

Cooter, R. and Luckin, B. (1997) Accidents in history: an introduction. In R. Cooter and B. Luckin (eds), *Accidents in History: Injuries, Fatalities and Social Relations*, pp. 1–16. Editions Rodopi B.V., Amsterdam.

CSCS (2015) *Construction Skills Certification Scheme* [Online]. Available: http://www.cscs.uk.com/ [25 September 2015].

Dekker, S. (2006) *The Field Guide to Understanding Human Error*. Ashgate, Aldershot.

Dekker, S. (2011) The criminalisation of human error in aviation and healthcare: a review, *Safety Science*, **49**(2), 121–7.

Donaghy, R. (2009) *One Death is too Many – Inquiry into the Underlying Causes of Construction Fatal Accidents*. The Stationery Office, Norwich.

Hare, B. and Cameron, I. (2010) Site manager training and safety performance. In P. Barrett, D. Amaratunga, R. Haigh, K. Keraminiyage and C. Pathirage (eds), *Proceedings of the CIB World Building Congress: Building a Better World*. CIB, Rotterdam.

Hartley, R., Finneran, A., Cheyne, A. and Gibb, A. (2011) Health and safety communication at Olympic Park – emerging findings. In *Proceedings of the CIB W099 Conference Prevention – Means to the End of Construction Injuries, Illnesses and Fatalities*. CIB, Rotterdam.

Health and Safety Executive (2003) *Health and Safety Regulation. A Short Guide* [Online]. Available: http://www.hse.gov.uk/pubns/hsc13.pdf [25 September 2015].

Health and Safety Executive (2006) *Successful Health and Safety Management, HSG65*. The Stationery Office, Norwich.

Health and Safety Executive (2007) *Managing Health and Safety in Construction – Construction (Design and Management) Regulations 2007 (CDM) Approved Code of Practice*. HSE Books, Suffolk.

Health and Safety Executive (2009a) *Underlying Causes of Construction Fatal Accidents – A Comprehensive Review of Recent Work to Consolidate and Summarise Existing Knowledge*. Phase 1 Report. The Stationery Office, Norwich.

Health and Safety Executive (2009b) *Underlying Causes of Construction Fatal Accidents – Review and Sample Analysis of Recent Construction Fatal Accidents*. Phase 2 Report. The Stationery Office, Norwich.

Health and Safety Executive (2009c) *Underlying Causes of Construction Fatal Accidents – External Research*. Phase 2 Report. The Stationery Office, Norwich.

Health and Safety Executive (2009d) *Health and Safety Induction for Smaller Construction Companies* [Online]. Available: http://www.hse.gov.uk/construction/induction.pdf [17 April 2015].

Health and Safety Executive (2013) *Managing for Health and Safety, HGS65*. The Stationery Office, Norwich.

Health and Safety Executive (2014) *Health and Safety in Construction in Great Britain, 2014* [Online]. Available: http://www.hse.gov.uk/statistics/industry/construction/construction.pdf [25 September 2014].

Health and Safety Executive (2015a) Construction Industry [Online]. Available: http://www.hse.gov.uk/statistics/industry/construction/index.htm [25 September 2015].

Health and Safety Executive (2015b) *Managing Health and Safety in Construction (Design and Management) Regulations 2015: Guidance on Regulations*. The Stationery Office, Norwich.

Hinze, J., Thurman, S. and Wehle, A. (2013) Leading indicators of construction safety performance. *Safety Science*, **51**(1), 23–8.

Hollnagel, E. (2004) *Barriers and Accident Prevention*. Ashgate: Aldershot.

Hollnagel, E. (2014) *Safety I and Safety II – The Past and Future of Safety Management*. Ashgate, Aldershot.

Hughes, P. and Ferrett, E. (2007) *Introduction to Health and Safety in Construction*. 2nd Edn. Butterworth-Heinemann, Oxford.

Independent (2014) Crossrail chief says worker who died in tunnel construction 'shouldn't have been where he was' [Online]. Available: http://www.independent.co.uk/news/uk/home-news/crossrail-chief-says-worker-who-died-in-tunnel-construction-shouldnt-have-been-where-he-was-9598314.html [25 September 2015].

Kamardeen, I. (2011) Web-based tool for affective safety training in construction. In C. Egbu and E.C.W. Lou (eds), *Proceedings of the 27th Annual ARCOM Conference*, pp. 309–18. Association of Researchers in Construction Management, Bristol.

Laitinen, H., Marjamäki, M. and Päivärinta, K. (1999) The validity of the TR safety observation method on building construction. *Accident Analysis and Prevention*, **315**, 463–72.

Lingard, H. (2004) First aid and preventive safety training: the case for an integrated approach. In S. Rowlinson (ed.), *Construction Safety Management Systems*, pp. 331–51. Spon Press, London.

Löwstedt, M. (2014) Taking off my glasses in order to see: exploring practice on a building site using self-reflective ethnography. In A. Raiden and E. Aboagye-Nimo (eds), *Proceedings of the 30th Annual ARCOM Conference*, pp. 247–56. Association of Researchers in Construction Management, Portsmouth.

Manu, P., Ankrah, N., Proverbs, D. and Suresh, S. (2010) The contribution of construction project features to accident causation and health and safety risk: a conceptual model. In C. Egbu (ed.), *Proceedings of the 26th Annual ARCOM Conference*, pp. 261–9. Association of Researchers in Construction Management, Leeds.

McEvoy, A.F. (1997) Working environments: an ecological approach to industrial health and safety. In R. Cooter and B. Luckin (eds), *Accidents in History: Injuries, Fatalities and Social Relations*, pp. 59–89. Editions Rodopi B.V., Amsterdam.

Ness, K. (2010) 'I know Brendan; he's a good lad': The evaluation of skill in the recruitment and selection of construction workers. In C. Egbu (ed.), *Proceedings of the 26th Annual ARCOM Conference*, pp. 543–52. Association of Researchers in Construction Management, Leeds.

Reason, J. (1990) *Human Error*. Cambridge University Press: Cambridge.

Richardson, S. (2006) How will this man make the Games safe for workers? *Building Magazine*, Delivering 2012 Supplement. November.

Rooke, J. and Clarke, L. (2005) Learning, knowledge and authority on site: a case study of safety practice. *Building Research and* Information, **33**(6), 561–70.

Spanswick, J. (2007) As near as dammit. *Building Magazine*. Issue 13.

Teo, E.A.L., Ling, F.Y.Y.L. and Ong, D.S.Y. (2005) Fostering safe work behaviour in workers at construction sites. *Engineering, Construction and Architectural Management*, **12**(4), 410–22.

Townsend, A.S. (2013) *Safety Can't Be Measured: An Evidence-based Approach to Improving Risk Reduction*. Gower, Farnham.

Warburton (1844) House of Commons Debate, 18 March 1844, vol. 73 cc1173-267, Hansard Online.

Whittingham, R.B. (2004) *The Blame Machine – Why Human Error Causes Accidents*. Butterworth-Heinemann, Oxford.

Chapter Five
Just a Bit Unsafe?

By implication there must be points at which 'suitable' becomes 'unsuitable'; there must be a point in the expenditure of effort to assess risk at which 'insufficient' becomes 'sufficient'.
Townsend 2013: 65

The language, or lexicon, of safety has had a considerable influence on the way we talk about it, and indeed are *able* to talk about it, on sites. In the UK, safety is very much grounded in the legislative framework, and so the way the legislation positions safety within management and individual duties and responsibilities will influence how safety is then translated onto sites, and how in turn it is used within work practice and daily operations. If these legislative roots of safety are unpacked, the way safety has emerged and developed over time can be revealed, alongside improved understanding of how this has influenced the ways in which safety works on sites, with considerations of safety itself, but also of unsafety and danger.

The Legislative Lexicon

Within the UK, legislation forms the foundations of many of the management systems and practices found on UK construction sites. Historically, this legislation was focused on danger, security

Unpacking Construction Site Safety, First Edition. Dr Fred Sherratt.
© 2016 John & Wiley Sons, Ltd. Published 2016 by John & Wiley Sons, Ltd.

and practice, but over time this has evolved into the much more complex 'safe' that now dominates contemporary law.

Early legislation was concerned with the mills and factories of the industrial revolution, and specifically the concern for the health and morals of the children working in them. The Factory Act of 1802 is generally considered the first 'health and safety act' in Britain (Putson 2013; Eaves 2014), although 'safety' is not actually mentioned or arguably even suggested in any form. Ideas of 'safety' did of course emerge within the legislation as it developed over time, but the language first used (in 1844, in just one clause of that Act) was actually that of 'security', and it was highly prescriptive – clearly stating that fencing of machinery was required, and this should be fitted securely, forming protection.

In fact, the term 'safety' only first appears in the Factory Act of 1878, in the heading given to Part II of the Act, although this section again plays a relatively small part in the whole legislation – there are only five clauses included within Part II (detailing fencing, violation on inspection, vats and structures, grindstones and cleaning of machinery in motion) – compared to much more considerable content addressing employment hours, mealtimes, holidays and education of children.

Within the 1878 Act, 'safety', or rather 'safe', makes its first appearance with relation to mill gearing:

> Cl 5 (3) every part of the mill gearing shall either be securely fenced or be in such position or of such construction as to be equally safe to every person employed in the factory as it would be if it were securely fenced.

Here, safe is positioned as a relative state, clearly given context in the form of secure fencing, providing a tangible example of what 'safe' should be in practice. This inclusion of a relative example is common in this Act, which also incorporates unsafety in the form of danger:

> Cl 6 any part of the machinery … is not securely fenced, and is so dangerous as to be likely to cause bodily injury to any person employed …

Again, an example is given in practice; the secure fencing – or rather any lack in the secure fencing – is positioned as the opposite to dangerous, rather than a more ambiguous 'safe' or 'safety'. Throughout this Act the use of mechanistic examples details the prescription of an unarticulated safety in practice, the more abstract, intangible concepts such as 'safe' only used to support more specific and detailed instructions. A very tangible and therefore readily understandable safety is constructed, both its presence and absence positioned within real-world activities and key aspects of the work itself. The workforce of this time probably had a pretty good idea of what safety was, clarified and set out in the legislation by what it looked (or ought to look) like.

The first combination of safety with health came in 1895, and again unsafety was explicitly addressed for both places and mechanistic processes:

> Cl 2 (1) … a process … cannot be so carried out without danger to health or to life or limb, by order … prohibit the place from being used … until such works have been executed as are in the opinion of the court necessary to remove the danger.

Unsafety remains a tangible measure – danger to life or limb – and also a tangible entity that can be removed accordingly. This construction of safety is again grounded in mechanistic and physical controls, fixed on concepts of *danger*. Safety is not something to be achieved; rather danger is something to be removed and safety as Hollnagel's (2014) idea of a non-event has been duly recognised. Identification of danger and its assessment has also been brought into the legislation, as a shared decision that can be made by the courts.

The twentieth century saw a significant increase in the number of regulations, and amendments to regulations, and also a change in the lexicon. The 1937 Act saw the introduction of some familiar, less prescriptive and much more ambiguous phraseology:

> Cl 26 Safe means of access and safe place of employment.
> (1) There shall, so far as is reasonably practicable, be provided and maintained safe means of access to every place at which any person has at any time to work.

This clause introduces the now familiar 'safe place of work' and 'safe means of access' of contemporary law, but this safety is different from that which has gone before. Here, safety has now become a standalone 'state', with no guidance provided for its attainment or what parameters would define its satisfaction. This is a significant shift. The 'danger' grounded evaluations and assessments sought by earlier legislation have changed perspective; this Act has refocused on safety, or more specifically safe as a state of being or place, and with just one clause as an exception to this rule within the Act, safety is now divorced from the realities of place or practice.

This emergence of safety as a descriptive and assessable state has repercussions for shared understandings of safety, as well as management actions for planning and control. A further shift was also made at this time through the introduction of the concept of 'reasonably practicable'. This is another addition to the legislative lexicon of safety, one which associates it closely with legal specifics and value judgements of risk and investment in terms of time, money and effort. This challenges previous prescriptive measures to counter 'danger', and indeed suggests a potential satisfaction to the very first arguments and debates that surrounded the introduction of safety legislation back in the times of the industrial revolution, allowing as it does the core values of production (time and cost) to also be applied to safety at work.

Both of these ideas, safety as a definable state and the notion of reasonably practicable, can also be found in the cornerstone of contemporary UK health and safety legislation: the 1974 *Health and Safety at Work etc. Act*. This Act, now the foundation of UK safety law, was developed from the examination of safety within UK workplaces as articulated in the Robens Report (1972). This Report took a highly progressive approach, and suggested that negative regulation and prescriptive legislation was not the best answer to modern safety management; rather risk should be managed by those who create it. In enshrining the Robens Report into law, the resultant 1974 Act drew on elements of established legislation, but also made one other significant change to the lexicon of safety. The Preliminary of the 1974 Act states its purpose:

The provisions of this Part shall have effect with a view to—

(a) securing the health, safety and welfare of persons at work;
(b) protecting persons other than persons at work against risks to health or safety arising out of or in connection with the activities of persons at work;

drawing on the construction of safety as a state of being, as employed in the 1937 Act. However, within Part (b) the Act also makes recourse to notions of danger, but in a different way to earlier Acts, which positioned danger in relation to practice, and provided examples of remedial action. In losing its prescription, safety has also arguably lost vital context, the focus on danger and the recognition of safety as a non-event has been put to one side. Instead, safety has been left to stand alone, and as a result has now taken on the position of a tangible 'entity' – something that can itself be 'at risk'. This has several repercussions which are explored in much more detail in Chapter 6, including consequential disassociations from practice, responsibility and ownership.

Throughout the General Duties of the 1974 Act, safety emerges either as a definable state, either of place, person or work, or as a tangible entity that can be ensured. The phrase 'reasonably practicable' is also evident throughout the Act; judgements of time and money have become a central part of the lexicon of safety. Unsurprisingly, the lexicon as enshrined in the 1974 Act has seeped into the many Regulations that it subsequently enabled. For example, within the Management of Health and Safety at Work Regulations 1999 that forms the basis for risk assessment:

> Cl 3. (1) Every employer shall make a suitable and sufficient assessment of—the risks to the health and safety of his employees to which they are exposed whilst they are at work.

Here, assessment is made of risks to the safety as a state of the person.

Also, within the Confined Spaces Regulations 1997:

> Cl 4 (2) No person at work shall enter or carry out any work in … a confined space otherwise than in accordance with a system of work which, in relation to any relevant specified risks, renders that work safe and without risks to health.

Although here, the state is the state of work itself, rather than person or place.

Interestingly, within construction industry specific regulations, whilst safety as a state can be readily identified, there is also recourse to earlier constructions of safety, more focused around danger and practice.

For example, throughout Part IV (General Requirements for All Construction Sites) of the Construction (Design and Management) Regulations 2015:

Safe places of construction work
Cl 17 (1) There must, so far as is reasonably practicable, be suitable and sufficient safe access to and egress from—

    (a) every construction site to every other place provided for the use of any person whilst at work; and
    (b) every place construction work is being carried out to every other place to which workers have access within a construction site.

Although bolstered by the value judgements of reasonably practicable, the state of safety for access and egress is here supplemented by 'suitable and sufficient'. This is an echo of earlier legislation – perhaps not as prescriptive as it once was, but certainly a more detailed construction of safe than that found within the 1974 Act itself (see Cl2.2(d)).

Construction as a relatively heavily legislated industry has also regained some of the earlier manifestations in practice around safety, potentially enhancing understandings of safety within the legislative context:

Excavations
Cl 22. (1) All practicable steps must be taken to prevent danger to any person, including, where necessary, the provision of supports or battering, to ensure that—

    (a) no excavation or part of an excavation collapses;
    (b) no material forming the walls or roof of, or adjacent to, any excavation is dislodged or falls; and,
    (c) no person is buried or trapped in an excavation by material which is dislodged or falls.

This clause is highly reminiscent of those found within the early Factory Acts, and notably safety does not even get a mention. Rather, focus has returned to earlier approaches and notions of unsafety: it is danger that is prominent, and it is the prevention of this danger rather than the establishment of 'safe' that is sought. This is very different from legislation that simply seeks an undefined 'safe' within work practices, which necessarily has to draw on different associations in terms of experience and understanding of those putting it into practice. Danger here is also positioned firmly within a context of work activities, the potential danger is spelt out in detail, and examples of how to 'prevent danger' in practice are included, notably through batters and supports. This is much more prescriptive and detailed than the simple 'safe place' of work found within the 1974 Act.

This is, of course, in part reflective of the fact that the 1974 Act attempts to be relevant to *all* workplaces in the UK, whilst construction regulations have the luxury of a very specific context. However, people generally do not unpick and unpack legislation to the level of detail explored here; it is likely that more general understandings and wider shared acceptances will have the most influence, and so are arguably the most important. Therefore whilst the relative familiarity with such specific and more general regulations will vary, it is certainly arguable that the dominant lexicon in many parts simply reflects and reverts to the 1974 keystone of UK safety legislation.

These shifts from the prescriptive to the abstract, the inclusion of value judgements, and perhaps most significantly, the change in perspective from danger to safety, will all have influenced the social construction of safety in the UK. Within the construction industry, it is this broader lexicon that has informed our management systems, which seek to put such legislation into practice, and therefore defined and framed how we *can* talk about safety on sites, which in turn has had significant influence on its positioning in practice.

Safe/Unsafe

It is not hard to find this legislative lexicon of safety – the legalese – on construction sites, often within safety documentation and induction materials. The safe 'systems of work' and 'working environment' found in Clauses 2(a) and 2(d) of the Health and Safety at Work etc. Act 1974, respectively can be easily found throughout our safety documentation. They are often either associated with general management practices or focused on specific work tasks such as falls/ fall prevention or excavations, where guidance often makes explicit reference to a 'safe system of work'.

This seeping of legalese throughout construction site safety is only to be expected. The site documentation will form part of an organisation's safety management system, and as such will mirror the necessary requirements needed to maintain the legality of its operations. Its prevalence in management rhetoric is therefore unsurprising, and its inclusion also legitimises the documents in which it is contained; explicit positioning of the legalese validating the way an organisation is presenting its approach to safety, robustly grounding it in the law. As a result, 'safe systems of work' and 'safe places of work' are common phrases of site life, and as such we all 'know what they mean'. Yet within a socially constructed world, such consistent agreement and consideration of contextual practice is certainly not something that can be so readily assured.

And indeed this lexicon does have consequences in its application to practice. The development of legislation over time saw the shift from ideas of 'danger' to those of 'safe', and so rather than seeking to prevent danger or remove risk, the lexicon has redirected the construction of safety to that of a descriptive state. Yet this also inevitably creates a reality in which 'safe' should be readily and indeed easily identified as such; something is either safe or it is not – but this does not allow for any middle ground. The contemporary lexicon of the law has had significant influence on our version of safety, and it has been polarised into just two terms:

Safe
Unsafe

This can be considered highly beneficial from a management perspective: the clear, crisp, binary approach to safety does facilitate management and assessment, as required by the legislation. It provides just two boxes to choose between when making the tick on the site inspection clipboard, and constructs safety as an easily documentable definition – a state that can be achieved, which in turn reconfirms 'safe' as something that can then be managed. Safety in this polarised state is able to meet quantifiable criteria, and can be used to deliver and help 'define' safety, allowing it to be benchmarked and measured.

This is found in ideas of monitoring and audits that can form part of a safety management system (Construction Skills 2010). The Construction Industry Training Board (Construction Skills 2010: 18) advises that safety inspections should be carried out by someone with the:

knowledge to be able to detect unsafe situations …

Such inspections must operate within a reality where something is either safe or unsafe, but this does not leave room for anything in between.

Polarised safety is also used within workforce interaction on sites, for example within induction slides we often say:

If it is not safe, do not do it.

Which again suggests there are just two identifiable states of safety – safe or not safe – and also assumes they can be easily identified and understood by workers. Within the context of inductions or training, such statements are often likened to work practice through the use of scenarios and associations with both personalised and active work tasks, yet the approach often remains firmly grounded in a static and polarised safety. But this again means there are no grey areas, nor is any development or change from one state, safe to unsafe or the other way around, considered or allowed for.

But that is precisely how construction sites *are*: things change. They are very different workplaces to any other, we create our own work environments as we go – if change didn't happen we would be out of a job! But this means that the same barrier around an excavation in the ground may be safe one day and unsafe the next; changes in ground conditions, the weather, the excavation process, all of these contribute to an ever changing place of work, very different from other industrial contexts. With situations ever changing, conditions can therefore easily flow back and forth along a continuum between these two polarised states, and so catching it at the moment of safe/unsafe is rather challenging – if not impossible. Should something nudge into the realms of 'just a bit unsafe', what action should be taken? And when elements of the 'unsafe' can be identified in almost any construction work situation, is the knowledge to identify them in a safety inspection actually of any practical use?

It is therefore unsurprising that away from the formal manifestations of legal requirements in safety management systems and documents, safety becomes much more fluid and flexible. When those who work on sites everyday try to adopt the polarised lexicon of safety with their own understandings of construction site life, it doesn't quite work. Evidence of this can be readily found in site-produced safety documentation (as opposed to formal corporate documents distributed by contractor head offices), such as induction slides, where the site team often try to reposition safe/unsafe within a wider context; examples of practice are given as they were in the earliest legislation, the acknowledgement of variation in individual judgements and assessments is made, and the ideas that people can be 'not sure about something' are now included and shared.

Such site-produced material often also contains explicit inclusion of change within understandings of safety:

Stop working if unsafe or unhealthy conditions develop.

Through a close association of safety to practice, this statement from an induction booklet has now positioned safety as integral to action, both consequential and preceding. This acknowledges that the construction site environment is a place of change, where unsafe or unhealthy conditions can 'develop'. Such change is not considered exceptional, although it does still remain an 'if' rather than a 'when', constructing safety, or rather unsafety, as a *potentially* developing process. Contrary to the polarised dichotomy of safety in the states of safe/unsafe, here safety is fluid, the changing environment itself the influence to safe or unsafe conditions.

Despite the dominance of safe/unsafe within corporate literature and management systems, some documentation does seek to reposition safety away from this polarised construct within the realm of management. For example, the Health and Safety Executive HSG150 guidance document *Health and Safety in Construction, HSG150* (2006: 21) acknowledges the complex and ever-changing fluidity of safety, by the need for:

maintaining healthy and safe conditions

Here, safety is considered as an ongoing process, it is still a state—but one receptive of change, acknowledged through the need for 'maintenance' within the work environment. The fluidity and flexibility of the work environment it is operating in has been embedded within this particular construction of safety.

When people talk about safety, its fluidity immediately emerges:

If you can get away with doing something slightly unsafe … it gets done quicker.

Here this subcontractor's supervisor has created another state of safety; something just 'slightly unsafe'. It has also been assessed and balanced against productivity, speed providing justification for the action. For this worker, safety does not exist in a definitive state, it is something far more fluid and flexible and dependent on the associated circumstances of the work being done. The close association of safety with site practice suggests that it is reflecting the variability of its context, and so safety becomes something that is fully accepting of and indeed develops alongside the changes inherent in the site environment.

Other understandings also seek to provide a more flexible measure of safety, through close interaction with site practice:

I work quite safe anyway.

Although inherently bound up with work, safety is again constructed by this supervisor as a more relative assessment. And again, rather than any firmer or higher commitment to a definition of 'safe' itself, for this supervisor safety is something that you do, rather than something that can be defined apart from work itself. When talking about his operatives, a similar relative 'measure' of safety is provided; whilst the supervisor initially emphasised that they were fully compliant with safety on the site where they were working, this became softened to:

They don't really break the rules.

Again this is associated with a level of flexibility within the reality of accepted behaviours, as well as the linked understandings of safety. Safety is not just one thing or the other, but something far more relative and fluid, and one that can be repositioned depending on its contexts.

This flexible and varied safety of the site is at odds with the formal safe/unsafe of many management systems. However, the need to ensure legal compliance seems to remain at the fore, and has consequently led to a relatively unchecked cascading of the necessarily restrictive lexicon of safety from the legislation into the formal site safety management documentation. Although such systems are perhaps necessarily reflective of the legislation, it is the positioning and shared understandings of safety *within* the work context that are arguably critical – it is usually during work activities that people get hurt, not when they are reading a safety management systems manual.

Despite the necessity for safety management systems to reflect the legislation, it is arguable that without some acknowledgement of the potential dissonance between safe/unsafe and the actual lived experiences of safety on site, safety management and its interventions will struggle for relevance. For those working on sites understandings of safety reflect the variability of its context, fully acknowledging and developing alongside the inherent change of the construction site environment, something at odds with any simplistic tick box measure of safety.

Safety and Unsafety

The polarised safe/unsafe of the legislative lexicon has unsurprisingly become embedded in organisational safety management practices and safety management systems. Binary evaluations of safety are therefore also found at the heart of various safety management activities, such as risk assessments and site inspections, and so have considerable influence on the construction of safety in practice.

Consider carrying out a site inspection. Whilst it is relatively simple to identify the 'unsafe', there may simply be too many potential hazards and risks to ever confidently proclaim 'safe' in a given situation. Gravity is arguably one of the key hazards we have to deal with on construction sites, and as a result there is always the potential for slips, trips and falls within any task. When considered from this perspective – and the fact that this is one of the most common causes of minor accidents on UK construction sites – the limitations of the lexicon quickly become apparent. Who can confidently declare 'safe' when people can trip up on the flattest and cleanest of surfaces?

Even without such considerations, the polarised lexicon demands that those carrying out such inspections take responsibility for declaring *safe*, and given the potential consequences, this may just be too large a step for supervisors and managers to take. As a result a default position of *unsafe* dominates; there are too many complexities involved in constructing 'safety as safe' in practice.

Yet this does not quite harmonise with the challenges made of safety management systems (see Chapter 4) which sought a shift from the impractical management of safety as a non-event, to a refocus on risks and hazards. Rather than the emergence of risk and its management, the binary choice between safe and unsafe has simply developed a shared understanding that *everything* is unsafe. This also aligns with another familiar hook also explored in Chapter 4: the seemingly unshakable notion that accidents *will* happen.

And as result it is often stated that construction is 'inherently dangerous', that our sites just *are* unsafe. This historically accepted 'truth' underpins shared understandings of safety on sites, and the way in which safety works in practice is not helping to support any challenge to the contrary.

For example, the manifestation of this 'truth' often begins at the very gates of the site, as shown in Figure 5.1. Construction sites label themselves 'Danger' as soon as the hoarding goes up, making danger something inherent in the very presence and existence of the construction site itself. Although such warnings are likely directed at those for whom sites are not a place of work, it still serves to create a reality of the construction site place as one of danger. Whilst safety is also frequently constructed at the gates of the site for the workers through a myriad of multi-coloured signs making associations with personal protective equipment (PPE), good practice or the promotion of safety programmes themselves, the signs that label 'danger' remain prominent. This paradoxical labelling of construction sites as dangerous only serves to further perpetuate unsafety within an environment where attempts are being made to actually seek the utmost in terms of safety.

Figure 5.1 Danger – Construction Site

This labelling of unsafety is not just found on the gates, but continues to regularly appear on the sites themselves. The ways in which we signpost safety and danger on our sites could be causing us some significant problems. For example, Figure 5.2 shows a common sign on sites.

The sign has a readily definable function, that is a warning of an immediate hazard, and is talking to its audiences through a familiar textual construction: 'Danger – Deep Excavation'. The location of the sign is of interest here. It has been attached to the perimeter barrier of the excavation, which confirms the immediacy of the danger. This location also suggests that a primary function of such signs is making a physical connection for the audience – that a deep excavation is a potential hazard, that there is 'danger'. The sign therefore justified the physical presence of the barrier, and why the audience are on a certain side of it. A further lesser function can also be drawn out: the identification of the location of the excavation works on the site, which would also likely have been apparent by ongoing works in the vicinity. Such signs are common on sites, concerned with the construction of danger around a specific hazard, through both their texts and physical location.

Figure 5.2 Danger – Deep Excavation

However, whilst 'danger' would suggest imminent or potential exposure to risk or harm, this sign is actually constructing 'danger' on the physical manifestation of 'safety' put in place to remove exposure to harm and unsafety. The pedestrian barriers to the excavation are a safety measure, yet have been labelled 'danger' by the signs. And this is a common practice. We barrier off and label, we fix down plywood covers and label, we lock doors and label. Yet what we are doing is labelling safety as danger, and such signage does not therefore construct the *safety* of the site, rather they construct *danger* around the very manifestations of safety in practice.

This would not be so problematic, if it weren't for the fact that we also label 'danger' as 'danger'. For example, another common site sign can be seen in Figure 5.3.

Although the sign makes the same opening gambit as that made in Figure 5.2, it is actually operating within a very different reality. In this instance the physical danger *does* remain, as the construction practice to which it refers is still ongoing in action: 'Men Working Overhead'. No physical barrier is provided, no other

Figure 5.3 Danger – Men Working Overhead

method of segregation used – here danger has been labelled precisely what it is – unsafety. In Figure 5.3, safety has not actually been implemented through any physical segregation of the area, and the sign itself is actually performing a very different function to that of Figure 5.2, despite employing an identical textual structure. As a result, a homogenisation has occurred within the labelling and consequential understandings of safety and unsafety – around circumstances that are very different in practice.

Safety is constructed as danger, but danger is also constructed as danger. We have signs that construct danger around safety in practice, and signs that construct danger around work in practice. The latter creates a reality of inherent hazards, supporting notions of an inevitable unsafety within the site environment as they are either tolerated or accepted as inevitable within safety management. The former creates hazards where they have actually been neutralised and controlled through safety management practice, creating danger around the activities of safety.

This clearly has significant repercussions for those working in such environments, where indiscriminate use of the labels of danger and unsafety, whether a situation is safe or unsafe in practice, has contributed to the homogenisation of sites into a place of danger – despite the efforts and implementations of safety management in practice to control it. This has the potential to lead to a shared level of 'danger fatigue': if everything is dangerous, even when safe, then tolerance and even ignorance of danger when it does manifest, whether it is labelled as such or not, is likely to be affected. It could be suggested that the construction site is crying wolf – in assigning 'safety' the label of 'danger', what resources can be drawn upon to identify the dangerous wolf when he actually does appear?

Signage also suffers from problems of physicality. It is often fixed and then left, becoming rapidly obsolete – the sign in Figure 5.4 suggests a deep excavation on the fourth floor of a car park building. This creates further complexity, and further potential for 'signage fatigue' – what signs should be paid attention to?

Possibly the inclusion of more information would negate some of the inconsistencies noted above. Rarely do 'Danger' or 'Caution' signs actually explain why events or activities are hazardous within their texts. Nor do they necessarily incorporate any statements of desired action from the audience, other than occasions when PPE is demanded – as discussed in Chapter 4. Often the workforce audience is left to determine what action should be taken as a

Figure 5.4 Deep Excavation?

consequence of the sign. Yet this lack of supporting information could simply be a consequence of the medium; signs are limited by their physical constraints which dictate the volume of content to be read at the necessary distance. Furthermore, many signs with reference to immediate hazards are informal; they are produced in site offices by site teams for employment in specific locations and at specific instances. This can result in differences of approach and assumptions of audience awareness and knowledge. A more focused and prioritised approach to the content, positioning and management of signage on sites could arguably help develop a more coherent understanding of, and differentiation between, safety and unsafety.

However, it must be remembered that this signage is born of safety management systems, and as previously noted this will inevitably involve attempts to bring together polarised understandings of safety with the more fluid safety of site realities. As a result *change* must be acknowledged as inherent within this environment, and the role it will play within these constructions of safety and danger. Whilst the sign in Figure 5.4 is no longer relevant and therefore has influence in negating safety in one way, it is also highly illustrative of the fact that within the construction site environment few things (other than the construction itself) are fixed. Therefore by labelling safety as danger, a 'safety net' is in place should things change. Whilst the first act of the sign is to seize the audience's attention, it is also able to advise of consequences should the environment change – should the plywood cover be shifted from the manhole below, should heavy rain destabilise the excavation batters – and should danger appear if the manifestation of safety is altered in some way. Again the problems of polarity emerge, and we cannot confidently proclaim safe hereafter. As a result, some of the most prominent constructions of safety within the site environment, the brightly coloured safety signage, are further contributing to the shared understanding that there is no attainable, ultimate 'safe' – instead we are operating in a sphere of constant danger, and therefore, of course, one in which accidents will indeed happen.

Safety and Risk

Risk and risk assessment are one of the fundamental ways safety and unsafety is managed on UK construction sites, and are prescribed by law. The Management of Health and Safety at Work Regulations (1999) require that risks associated with any work activity are assessed before work starts. Risk assessments form the basis of many Health and Safety Regulations, and a standardised format and prescribed approach to the process often forms part of the safety management system toolkit in the establishment of 'safe systems of work'.

Fundamentally, the process of risk assessment arguably segregates it from the selection of the work methods. Although it should be a driving factor in making such decisions, often time and money outweigh safety considerations. As a result risk assessment is often applied after the event to construction activities pre-determined through design, despite the best efforts of the Construction (Design and Management) Regulations (2015), or simply common and familiar work practices. Risk assessment can simply become an add-on to the work itself. Indeed, the use of risk assessments in practice has been strongly criticised; the Health and Safety Executive (2003) has regarded them at times as of little value, generic, and including no operative consultation. In the UK risk assessments can also be problematic due to issues of literacy, as well as a multinational workforce who may not speak English at all.

In the UK, risk assessments are also arguably one of the most badly executed of our common safety management activities. Rather than a process for active involvement and engagement with decision making and planning, they are often seen as a paperwork exercise, a simple 'cut and paste' of the risk assessment documents between projects as the supply chain moves around from site to site (forgetting to change the project and main contractor's name on the top of the risk assessment documents is often the tell-tale mistake). It can of course be argued that many hazards and risks do remain the same – and indeed some will; for example good house-keeping of the work area is a necessity in all construction tasks to avoid slips, trips and falls. However, in many other cases the work is *not* the same, and when hazards are included that do not exist and hazards are missed off that do, the whole risk assessment process becomes worthless.

In addition, the fact that most skilled operatives are trained to carry out one trade or type of construction work leads to repetition of tasks from site to site, which can in turn lead to complacency. The task has been undertaken a hundred times before and as a result it is not uncommon for operatives to happily sign unread risk assessments, which then simply sit in a folder in the office, with no influence on safety on the site, again serving to reinforce the idea that safety can be readily 'signed for' and then dismissed. The paperwork all too often becomes the focus, forming an unhelpful and bureaucratic part of a safety management system. As a result hazards and risks are themselves not particularly welcomed within the construction industry, and there is a shared derision of the management process, and indeed the risks themselves.

Indeed, even when the approach to risk assessment is more proactive and detailed, concern has been raised by the Health and Safety Executive (2013:12), that the:

> focus becomes the process of the system itself rather than actually controlling risks.

We may well carefully document our hazards and risks, and detail out the mitigation and control measures required for safety, but do not seem to be as good at putting this into practice within the complex, changeable site environment.

This has therefore contributed to a shared understanding of risk within the construction site environment as something time-consuming and irrelevant. This is reinforced by the continued prioritisation of 'safety' on sites through safety management systems as explored in Chapter 4, and as a result the constructs of risks and hazards further diminish in the dialogue.

A disassociation of safety and risk can also be identified from the alternative perspective. In the Health and Safety Executive's (2014) leaflet, titled *Risk Assessment – A Brief Guide to Controlling Risks in the Workplace*, 'safety' is only mentioned four times – five if you include its presence in the Health and Safety Executive's own name. Safety is alternatively associated with responsibility, representatives, improvement and as something owned by businesses, the tangible safety of management coming to the fore. But this is only peripheral contextual framing; far more prominent are risks, hazards, injury and ill health – arguably the important practical aspects of 'safety'.

However, both of these contradictory positions – safety as the dominant discourse with risks and hazards reduced to bureaucratic 'noise' and risk as the dominant discourse itself disassociated from safety in practice – have arguably limited the inclusion and integration of hazard and risk within wider understandings of safety. Despite the prominence of hazards and risk within guidance documentation and (ideally) practice, they are not always so coherently partnered with, and positioned alongside, 'safety' on sites.

A further characteristic of risk that should be examined here is the familiar construct of 'risk tolerance' that is often applied to the construction industry. Indeed, risk itself has a special place in the construction industry – and it is actually something we do well, or not, depending on your perspectives. Whilst most people take some risks with their safety every day, be it driving a little too fast or crossing the road in too small a gap, in construction those risks are slightly different. Despite happily taking such everyday risks, most people would not want to climb a tower crane, or walk on a scaffold 25 stories high, or crawl a cradle up the side of a building hoping the wind doesn't change. But someone has to do it to get the job done, and so shared understandings of risk within construction have been developed within these specific parameters and contexts.

In its widest possible sense, the construction industry can easily be considered a risky operation, as explored in Chapter 2. Financial risks are taken in terms of developments, every tender is a risk in terms of the time and budget requirements, not to mention risks arising during the construction phase. From directors to steelfixers, a high level of risk tolerance permeates the industry. Everyone takes risks, albeit of a differing nature, and in part that is arguably what attracts people to the work. In an industry seen as macho and dangerous, studies have shown that many of the people attracted to work in construction show these characteristics in a much greater proportion than those who work in more grounded occupations, for example selling cars (Whitfield 1994). The construction workforce likes taking risks: it likes climbing higher and tunnelling deeper than most people would ever feel comfortable with.

Indeed, within this potentially risk-nurturing context, the opportunity for risk taking and facing danger could even have been precisely *why* operatives entered the industry originally. Taking a job in roofing, scaffolding or cladding is far more exciting than just sitting at a desk. Construction work undeniably requires a higher level of 'risk tolerance' and risk-taking behaviour than the average occupation (Cooper and Cotton 2000), and as a result shared understandings around an acceptance of risk have become commonplace. The segregation of risk from 'safety' within safety management systems, its derision through the risk assessment process, and the context in which these management process and practices are operating, has meant that risk has just become an inherent part of construction site life.

However, a level of risk tolerance is permitted, even encouraged to the level needed to complete the work, but *then* a line is drawn. This line may be positioned to reflect the level of 'risk tolerance' as dictated by legislation, safety management systems, construction site rules or risk assessments themselves. Yet understandings of risk often struggle to 'fit' within what is essentially an arbitrary framework, and so safety does not always 'work' alongside risk – safety becoming just one more constraining aspect to the wider risk tolerance found within the industry itself. That people may actually *like* to take risks does not sit well with safety management systems, which align with management the theories of risk compensation that state that everyone has a risk thermostat which should be set firmly to zero (Adams 2006). Safety management systems struggle to account for deliberate risk taking, it becomes something abnormal (Douglas 1992), despite the way it is constructed and embedded in shared understandings and contexts throughout our industry.

The following stories tell how six operatives recollect their participation in an event that led to their disciplining for a safety violation on site, and in doing so they also share their understandings of risk.

The Tree Surgeon's Story

We had to reduce a big poplar tree on this site by 20%. The safety guy from the contractor we were working for came down at the start of the job to check risk assessments, that we'd put cones out and the like, crossing all the 't's. He was concerned we didn't have adequate ear protection for the chipper, just for the chainsaws and so on. But in my opinion the biggest risk was the tree. They'd let it grow too much, it had a 70 foot spread with limbs sheared out, all long thin branches. They'd have been better pollarding it, just cutting it off to the trunk and letting it grow again, but they didn't want that. I had to go out on those thin branches to reduce it. I think we took a risk with that job, poplar's brittle, you don't know what it's going to do, it would have been safer to pollard it. We later found out that they'd asked another company to do it, but they'd insisted on using a MEWP [Mobile Elevating Work Platform], which would have been the best way, but the contractor didn't want to pay for it. They asked if it could be done without, and yes it could, but it wasn't the best or safest way. They come on site worried you've not got a high-viz on when there's no one else around, but want you to take a risk with a tree to save a few quid. It's all paper safety these days.

The Cable Puller's Story

We were pulling the mains along the corridors, the kind of cable most people can't even lift, never mind bend round corners. It was all open ceilings with all the trays and pipes on show, and we were at this corner and the tray we're laying in is right over next to the wall and the gap you're supposed to get in is right in the middle. So you've no chance of reaching and bending through this hole to get the cable down in the tray and clipped on. No chance. We had a little MEWP, a corridor sized one, but that wouldn't fit up the gap, it was too small, so

I climbed out and up into a couple of the other trays to get this cable in. That's when the site supervisor saw me and ordered me down. The trays were fixed on, they would hold me, I'm not that big, but the supervisor wanted to know how I knew they would hold? Did I fix them? Well, no, but we needed to get the job done. The supervisor did explain they'd only stopped me because they didn't want me getting hurt, and we agreed that designers didn't know what they were doing. Why put that cable furthest from the opening? Why design the ceiling so it's too small for anything to get up and reach across other than your head and one arm? They should try fitting some of this stuff, might make them realise. Anyway, the site supervisor went and got the M&E supervisor and sent us out for a smoke, we were getting really cheesed off, we'd been struggling on this job all along and people kept stopping us, how were we supposed to make our money? When we came back the M&E supervisor had agreed to take some of the other trays out so we could get the MEWP in properly. To be fair, it didn't take too long to sort and the site supervisor was just looking out for us in the end. But things should be better designed and thought through properly.

The Window Fitter's story

We were fitting window frames on the second floor of the building off an external scissor lift. All the staircases were blocked so there wasn't any access in the building. We were under pressure to get the job done, both from the contractor and we were on price. In order to get inside, I lifted the scissor up to the window level and used the door to climb in through the window. It wasn't ideal, but I took action and thought it was safe. There was a hop up inside to climb down onto and the scissor was only an inch, if that, off the building. I couldn't have fallen. I eliminated the risk. But one of the managers saw us and I got a verbal warning. I'm not in the practice of doing stupid things. There's a risk in every job and I minimise it as best I can. Safety has gone too far, it doesn't allow you to do your job any more.

The General Operative's Story

We needed to fit a closure piece into the curtain wall movement joint about 7m off the ground. There had been a scaffold there which I could have easily reached off, but the top had already been struck, leaving only a lower platform. To reach up we lifted a podium step up onto the scaffold and then I had to stand on the handrails of this to reach the joint. What I was doing didn't really come into it. It needed doing, and it needed doing then, I suppose I shouldn't have been doing it, but it was a two minute job and with the cost and time to put the scaffold back up, and it would have delayed the floorers too. That platform had to come down that day anyway, there wasn't time, one of my mates was footing me, it was fine but just looked bad. It wasn't the right way to do it but it was the easiest. It was my fault it needed doing like that anyway as I missed the scaffold strike, I should've done it the day before. I was probably most afraid of getting caught.

The Electrician's Story

I was doing some work up a stairwell, so I footed two ladders at the bottom, placed across each other so I could put a scaffold plank across to make a platform. I was quite happy working off it, they were both my ladders, class one, and I made sure I chose a very good board! I knew it wasn't going anywhere, anyone with a grasp of physics would've known it wasn't going anywhere. But the client on this job had his own health and safety inspectors, they'd come and do spot checks, and they caught me. They insisted I get a proper scaffold, which I did and charged the client because I hadn't priced for it, all to do the 10 minute job I was up there for. I know it wasn't a proper platform, I could step off in theory, they have to assume there are some idiots on site I guess, and if I'd dropped anything I hadn't cordoned it off, although you would've known I was there, you couldn't get up the stairs with me in the way. I do appreciate it, but you make your own decisions at the end of the day.

The Roofer's Story

We were fitting the flashings to this kalzip roof out of a scissor. When we got round to the next elevation, I saw we'd left a lanyard on the roof. We were always getting a bollocking for not clearing up, so I just went onto the roof to grab it. The engineer saw me and went mental. I got thrown off site for the day for doing that. For tidying up! I wasn't going to fall, I suppose I might have slipped, but I wasn't going to, I'm on that stuff all the time, you know how to walk on it. I know you're not supposed to get out of the scissors, go from one to the other like, but it got it done there and then. I suppose it was fair enough, it's the management's job at the end of the day, but still, I was alright, I knew what I was doing. I wouldn't do it again, not when anyone could see at least. Lost me a day's money that did.

All these stories tell of a shared acceptance of risk within the construction industry workforce. In four of the stories, risk does not emerge, nor is it labelled as such by the operatives. This was not risk; it was just what they do every day, nothing dangerous, unsafe, or even out of the ordinary. For the window fitter, although risk was articulated, it was also negated and considered eliminated, despite the fact that the behaviour was deemed unsafe by site management. The tree surgeon's story is irregular to this pattern, but although risk did emerge within this story, positioned within a context of the individual's own specialist knowledge and skills, it was still negated within the work practice.

However, in all the stories, the operatives positioned and labelled their violations, for a variety of different reasons, to be 'not right' in the wider construction of 'correct' work methods and safe behaviours. Only in one case study (that of the tree surgeon) were the labels of 'risk', 'dangerous' or 'unsafe' used by an operative to describe their actions in that instance. What are deemed as risk-taking behaviours by some, such as site management, are simply constructed by operatives as just part of the work; risk is simply not part of their lexicon of safety.

The positioning of the 'risks' within the wider site context can be found in two distinct constructs: those of a personal nature and those that are seemingly created by the 'management' – a term used to refer to anyone above operative level, but is also in part a reference to other aspects of the industry context. On a personal level, operatives place ideas of risk firmly within their wider under-standing of work, the timing of a task, the argument that 'it's a two minute job' used as justification for the action. As noted in Chapter 2, work on price clearly has influence in terms of the con-struction of risk, and means a delay due to safety is a delay that cannot be afforded; operatives are at work to earn their money as fast as possible. This is also strongly linked to the need to get the job done. Whether working on price or not, the need for productiv-ity is key in the reflections of risk-taking activities. Confirmation of autonomy is also evident – 'you make your own decisions', 'I took action' – which emphasises the independent role the workers played in their own unsafety. Yet frustration with 'management' was also evident and in certain cases positioned as the underlying cause for risk-taking behaviour. Through poor management of the work site, by applying these pressures for production, through the fundamentally poor design of the structure or simply to save money, 'management' are seen as additional actors and partici-pants in the individual's constructions of risk.

Both individuals and the construction site context are therefore highly influential in the social construction of risk on site. For some workers, personal benefits, both tangible in terms of money or abstract in terms of self-development and autonomy, are used to position and indeed justify risk within their own understandings. Occasionally the management structures in which they operate are drawn upon as support and validation of such understandings. Indeed the high tolerance to risk taking found within the construc-tion workforce could even be seen as indicative of the *require-ments* for the work. Construction is often seen as risky by the general populace, but is not considered a risk by operatives, even when acknowledged boundaries of safe working for the task are stretched or broken.

However, the shared understanding of 'safe' that can be drawn out from within these stories also implies that training and educa-tion are unlikely to provide a simple solution. This is simply reflec-tive of the complex and contradictory nature of people. Safety can be constructed as inherent in work practices, but just as readily unsafety can be constructed as part and parcel of construction site life, and nothing out of the ordinary. The way workers consider and position themselves within the framework of safety rules and regu-lations on sites is also closely linked to these understandings of unsafety and risk, and management of these through both engage-ment and enforcement are discussed in more detail in Chapter 7.

Summary

The role of risk is critical to understandings of safety in principle. Risk is something the construction industry understands all too well; it is valued and even needed within its workforce to get the job done, and so proves highly problematic when positioned alongside ideas of safety. The lack of consistent and valued associations between risks and safety, coupled with the restrictive polarised lexicon, have devalued the ways in which shared understandings of risk can contribute to safety, or rather unsafety.

Indeed, when safety is considered from the perspectives of unsafety or danger, several different realities emerge: there is the reality in which hazards have been resolved yet are still articulated as danger, the reality in which hazards are accepted and considered an inherent part of the site and there is the reality where hazards remain hidden and safety is articulated simply through the requirements of PPE, all of which further create a complex and inconsistent context for unsafety on sites.

As a consequence of this, when we talk about safety using our contemporary lexicon, often the question becomes one of 'how safe is safe?'. This creates a reality where safety in practice is highly fluid in terms of practical associations and levels of overall commitment. That safety is not 'easy' becomes rapidly apparent, and the challenges made to the simplistic polarised construct of legislation and management systems by those who use and work with it every day suggest a highly complex and contradictory nature. Importantly this results in a clear disassociation between aspects of policy and practice. Although formal documentation and legislation appear to consider safety as relatively simple and easily defined within the limited polarised scope of safe/unsafe, practice has necessarily unbent to allow for emergence and change; safety becoming a fluid, flexible and mutable state, reflecting the variability of its context, fully accepting and developing alongside the change inherent in the site environment.

This naturally has repercussions for the enforcement of safety. The legal framework provides management with requirements for implementation, grounded in the safe/unsafe lexicon, yet how can these labels be applied in a fluid and flexible context, where safety and unsafety emerge and dissipate on a regular basis? Where something can easily be, in the words of one site operative:

Just a bit unsafe

In such cases where should management draw their line of safety, beyond which violations will not be tolerated, when it often depends on a split second, and needs someone there to witness it? This closely links to both the enforcement of safety rules and also worker engagement with safety. This is explored in more detail in Chapter 7, and of course begins to explain why safety violations also form such an accepted aspect of site life.

However, the legislative lexicon has also introduced the idea of safety as a tangible 'entity' – even to something which can be 'at risk' in its own right, and something that can therefore impact both ownership and responsibility, and which can lead to the positioning of safety against work, but also work against safety.

Acknowledgements

This section is based on work previously published by the Association of Researchers in Construction Management:

Rawlinson, F. and Farrell, P. (2009) The vision of zero risk tolerance in craft workers and operatives: an unattainable goal? In A.R.J. Dainty (Ed.), *Proceedings of the 25th Annual ARCOM Conference*, 7–9 September 2009, Nottingham, Association of Researchers in Construction Management, Vol. **2**, 1203–12.

And in the journal, *Construction Research and Innovation*:

Rawlinson, F. and Farrell, P. (2010) 'But we like risk'. *Construction Research and Innovation*, **1**(1), 46–50.

And in *Construction Manager* journal:

Sherratt, F. (2013) CPD: site safety signage. *Construction Manager*, January.

References

Adams, J. (2006) *Risk*. Routledge, Oxford.
Construction Skills (2010) *Site Safety Simplified*. Construction Skills, King's Lynn.

Cooper, M. and Cotton, D. (2000) Safety training – a special case? *Journal of European Industrial Training*, **24**(9), 481–90.

Douglas, M. (1992) *Risk and Blame: Essays in Cultural Theory*. Routledge, London.

Eaves, D. (2014) *'Two Steps Forward, One Step Back': A Brief History of the Origins, Development and Implementation of Health and Safety Law in the United Kingdom, 1802–2014* [Online]. Available: http://bit.ly/1nJPvjJ [1 October 2014].

Health and Safety Executive (2003) *Causal Factors in Construction Accidents, RR156*. The Stationery Office, Norwich.

Health and Safety Executive (2006) *Health and Safety in Construction, HSG150*. The Stationery Office, Norwich.

Health and Safety Executive (2013) *Managing for Health and Safety, HGS65*. The Stationery Office, Norwich.

Health and Safety Executive (2014) *Risk Assessment – A Brief Guide to Controlling Risks in the Workplace* [Online]. Available: http://www.hse.gov.uk/pubns/indg163.pdf [8 October 2015].

Hollnagel, E. (2014) *Safety I and Safety II – The Past and Future of Safety Management*. Ashgate, Aldershot.

Putson, D. (2013) *Safe at Work? Ramazzini versus the Attack on Health and Safety*. Spokesman Books, Nottingham.

Robens, A. (1972) *Safety and Health at Work: Report of the Committee 1970–72*. Her Majesty's Stationery Office, London.

Townsend, A.S. (2013) *Safety Can't Be Measured: An Evidence-based Approach to Improving Risk Reduction*. Gower, Farnham.

Whitfield, J. (1994) *Conflicts in Construction: Avoiding, Managing, Resolving*. Macmillan, London.

Chapter Six
Safety versus Work and Work versus Safety

Progress is not the victim of health and safety regulation.
Putson 2013: 172

Alongside the construction of safety in its polarised forms of safe/ unsafe, the legislative lexicon also at times positions safety as a tangible 'entity' that can stand alone – something that can have things done to it, something that participates in work and, but also therefore, something that can be set aside, ignored or even dismissed. Chapter 4 saw how safety management guidance has recently sought to embed rather than segregate safety from work practices, but shared understandings of safety are not so easily influenced, and can take time to change. The legislative roots and indeed common ways of managing safety on site – through safety management systems for example – still influence and impact how safety and work fit together on sites. It is arguably at the site level of practice that safety becomes most relevant. It is on sites where shared understandings of safety come together, where safety is enacted and made manifest, and sadly where people actually get hurt.

The Non-Productive

The discussions of safety management systems in Chapter 4 were concerned with the *management* of safety within such approaches; which also necessarily develops the understanding that safety is

Unpacking Construction Site Safety, First Edition. Dr Fred Sherratt.
© 2016 John & Wiley Sons, Ltd. Published 2016 by John & Wiley Sons, Ltd.

something that *can* be managed. However, the use of a safety management system makes a different contribution if we refocus on the *safety* part of this label; and in doing so safety immediately becomes a 'thing' in its own right. As a direct consequence of this, safety is set apart from the practice of work itself, positioned alongside other non-productive aspects of industry.

Safety is constructed as an *entity*; it has become a tangible 'other' – an independent actor in the workplace. By seeking prioritisation of safety in principle through its own clearly defined management system, this approach has actually developed a significant rift between safety and the world of work. Safety has become disassociated; engagement or interaction with the site environment and its work practices has become unnecessary for its existence or function – the system, and therefore safety, can simply stand alone. This is in sharp contrast to the commonly articulated aims of safety management systems and industry safety programmes, which seek to instil safety within all aspects of the construction site environment, and try to embed safety principles within all work practices (Lingard and Rowlinson 2005; Health and Safety Executive 2007).

It could therefore be suggested that the common ways in which we have tried to manage safety have actually helped to hamper the process; in seeking to reflect the 'safety' of the legislation and explicitly address this aspect of construction work, safety through a safety management system has actually become segregated from the work itself. And this has not been helped by the common practice of making associations and amalgamations with other non-productive elements of construction work.

Indeed, the term safety management system is often the abbreviation for a:

Health and Safety Management System

The common amalgam of health and safety (H&S) reduced simply to 'safety' within the safety management system shorthand. It can and indeed has been suggested that the focus on the S and not the H&S has actually been to the detriment of occupational health – which was ironically at the heart of the earliest legislation to benefit workers of the emerging industrial age. It has often been argued that safety has been unduly prioritised in the construction industry over health – which is something far more damaging to the workforce when the numbers and nature of construction health complaints are examined in detail (Health and Safety Executive 2014).

There are essential differences between these two components; whilst safety is of the immediate in terms of time and place, health can take much more time to manifest and fully emerge. Consequently, their constructions within our social landscapes are very different, and therefore the considerations around their management should reflect this. That health is more often than not the neglected partner in this relationship, particularly in the construction industry, also indicates the potential difficulties in its management on sites. Consider dust for example, it has significant consequences in terms of lung diseases from continued exposure, yet how to effectively contain something that seems to gather in every corner of an unfinished building remains problematic. The poor levels of occupational health within the industry can be seen as evidence of its neglect, but may also be in part a direct consequence of this H&S amalgamation, which has seen the far more dramatic and impactful safety take precedence in terms of management prioritisation and control.

Although safety may have benefited from the development of management through safety management systems, arguably far more damaging is the bundling of safety with other, far more disassociated practices. H&S has expanded to incorporate other site management roles, and people are expected to be able to perform as, for example, a:

Health, Safety and Environment Advisor

The addition of environmental management to health and safety can be seen as a reflection of the increased prioritisation this aspect has under the sustainability agenda within the construction industry, or, more cynically, as the result of the need to stick its management *somewhere*.

Many construction organisations have now positioned this new amalgamation of HS&E under the umbrella of sustainability on their websites, or within the even broader remit of 'responsibility', corporate, social or otherwise (Rawlinson and Farrell 2010). Safety of the workforce has become a sustainability issue – without their longevity work cannot be carried out, and the construction industry should not be damaging one of its 'resources' in this way. The sustainability discourse is highly complex (and indeed far too complex to unpack here), yet this repositioning arguably develops associations with morality, the generosity of both corporate spirit and investment to safety, which is somewhat misleading when its management on sites is something that construction organisations simply *should* be doing by law. Putting safety under the banner of responsibility makes it vulnerable to the packaging and presentation of safety on sites – turning safety into a commodity for PR and marketing – rather than retaining focus on the fundamental methods and processes of its management in practice. This has the potential to even further segregate safety from the realms and realities of production.

In some instances safety has even seen the addition of quality to its various amalgamations, creating a:

Health, Safety, Environment and Quality Advisor

This is arguably the maximisation of the non-productive – including all the recognised parameters of construction industry production, with the notable exceptions of time and cost. How these aspects of construction work can all be bundled together effectively must be questioned – simply carrying out this role must be not only totally exhausting but also something of a challenge, when the knowledge and skills needed to provide detailed and comprehensive advice within such a wide scope is soberly reflected upon.

Yet safety is often absorbed into these various bundles, and the people assigned to carry out these roles are often seen on the project organogram off to one side, their only link being to the project manager. Whether this is intended to be reflective of their prioritisation within the project, or in fact segregation in the truest sense of the word, the end result is the same; separation from the operations team who get on with the *real* work of construction.

This is highly problematic when considered from a constructionist perspective. Health, safety, environment and quality management have now become one; they have come together to form an amalgamated mass of the non-productive. This potentially has several different consequences.

Firstly, this segregates safety from practice, constructing it as a disassociated entity that sits to one side on site, along with all the other construction management processes remote from the site activities that actually contribute to the more important (and money-making) business of productivity. Positioning safety in this way does not engage it with work; rather this reinforces the construction of safety as an add-on, something without tangible merit when considered within the real purpose of the project.

Secondly, bundling implicitly ascribes equal importance to the component elements, regardless of their individual relevance or significance in a given situation. This creates inherent associations between the component elements, despite the significant variation between them in terms of practice and associated behaviours or actions. As a result, the impact of each individual element on site operations is arguably reduced, ascribing all elements within the bundle an equal level of priority. Whilst health and safety may be distinct, they are at least still associated in terms of their focus on the individual. The incorporation of environment or quality to this pairing is far more incongruous, and positions unrelated and impersonal elements alongside those concerned with the impacts of work on the people actually carrying it out. Such bundles therefore construct unrealistic amalgamations in terms of the relative associations and implications of the contributory elements themselves.

Thirdly, bundling arguably reduces these individual elements to an acronym in the form of a convenient generic term, all too similar to the dismissive 'elf'n'safety' explored in Chapter 3. Indeed, the continued use of such bundles by organisations actively seeking to independently prioritise and effectively implement safety, health, environmental and quality management systems, may actually find their use to be one of the limiting factors to the inclusion and integration of these aspects within the principles of work on their sites.

As noted in Chapter 2, the non-productive aspects of construction work – those that are not time and money – can often struggle to impact on site operations. Fundamental problems of segregation from practice through the development of safety as a separate entity are further compounded by its inclusion in these non-productive bundles. Such terminology and management associations certainly influence how safety 'works' on sites, and unfortunately such approaches are inherently unable to support the statement that:

Safety is our number one priority!

and instead suggest something really quite different.

Segregation and Integration

Indeed, the influence of such acronyms and the use of safety management systems can be readily found in how people talk about safety within the context of its practice. For example, phrases like:

Safety has an important role to play

are often used on sites and can often be found in safety management material, inductions or training sessions. This entity of safety has the ability to act in practice and play a role, and an important one at that. But again this is in contrast to any positioning of safety as *inherent* within the workforce, their activities, or wider site operations; it is once again a tangible 'other' – an independent actor in the workplace. In terms of safety management, this throwaway phrase seeking prioritisation of safety in practice has actually developed a rift between the two – and one which has repercussions with regard to the acceptance and relative positioning of safety and work.

Establishing safety as an active participant in the site environment also gives it a unique 'identity' within this context; and inevitably this identity is distinct from the identities of the other social participants, i.e. the individuals that make up the site workforce. As a result, *responsibility* for safety has also shifted to this new entity, which potentially negates any significant responsibility or ownership of safety by the workforce, in terms of their understandings or actions; safety has actually been constructed as the key participant in its own manifestation.

This has significant consequences in terms of practice. If safety is understood to be a separate entity, although present on site it is not necessarily engaged, and is therefore likely to be kept apart from the more prominent considerations of production.

Although it could be suggested that positioning safety as a separate entity is a simple rhetorical manifestation of reference to an abstract concept, reflective of safety management systems, it is equally suggestible that it is *precisely* the associations with ownership and responsibility that are important here, and have actually helped develop our shared understandings. In constructing safety in this way, it is set apart from any personal responsibility, ownership or action and ultimately becomes either its own, or more likely within the context, someone else's responsibility.

This will also have been influenced by the polarisation of safety as explored in Chapter 5. The familiar labels of safe/unsafe may simply be too remote from the fluid and changeable environments of the site, which in turn has led to resistance in the development of a shared construction of safety that is fully integrated with practice, and instead resulted in its establishment outside of productive work in the form of an essentially excluded entity.

However, there are also other ways of talking about safety that do reflect the desires of safety management, and safety *can* be totally tied up with practice. Such constructions of safety quickly shift away from 'safety' itself and move towards talk of practice, social interaction and personal ownership, as this operative says:

> Safety, it's everything isn't it, not just for yourself but for everyone else around you, you've got to be looking out for everyone … if he's not looking out for me, and I'm not looking out for him, well, we're not going to be getting anywhere, are we?

Here, safety, although still abstract in terms of scope, is something immediately placed within a context of practice and social relationships between the members of the workforce. This operative has also taken ownership and personal responsibility for safety, and also sees that as a requirement in his fellow workers. Safety here is constructed as a team game, the operative is a part of that team, and there is the need for progress and action towards a team goal. This construction of safety, as inherent in practice, is precisely what legislation and safety management systems theoretically set out to do. In contrast to safety as entity, safety here is embedded within the actions and interactions of the site. This can be heard, seen and found within many different elements of site life, within a wide variety of specific work practices and processes as well as more general social interactions on sites.

But these different constructions of safety are also readily able to demonstrate that there *are* distinct and different understandings of safety to be found within the site environment. Indeed both can be enacted by the workforce at the same time – such is the nature of our social world and therefore the realities of our construction sites. Safety is either its own entity, separate from the individual in both action and responsibility, or it becomes an inherent part of social interaction, bound up with people and practice in terms of their own actions and responsibilities. And people often draw on either or both of these two different constructs within their interactions, even shifting between them within a single conversation; safety becomes an entity in some contexts, yet inherent within practice in others.

These contrasting constructions of safety serve to illustrate the complexities of safety itself, and also indicate the very personal nature of the associations and interpretations that ultimately inform interactions and responses to safety within the construction site context. Although it could be suggested that safety management systems have achieved some success in embedding safety within construction site practices, there is still the potential for safety to be constructed as somebody else's problem. This has repercussions for the engagement of the workforce with safety, something discussed in more detail in Chapter 7.

Problems of Production

One of the most readily apparent ways safety is segregated from work practice is through its associations with production, and this is nothing new. Since the first legislation to address safety and work was proposed in nineteenth-century London, counter arguments have been made that safety management would be a substantial imposition to the far more important considerations of production. As Mr Hyatt stated in the House of Commons back in 1833, the introduction of punishment for industrial accidents, and the need for legislative control of the work environments, were part of the significant:

> ... restrictions the House was called upon to impose on this great manufacturing country ...

And at the time this 'great manufacturing country' was often given a much higher profile within political debates than worker safety – or indeed any other considerations – and so began the development of the context for safety management legislation that is still familiar today. Arguments about the value of British exports and the threats of competition from foreign markets established an early dichotomy of safety versus productivity; the language of economics used against that of human welfare, despite their joint participation in the ongoing industrial revolution of the time. Although our world is now much changed, echoes of the past can still be heard today, and are even made manifest in the 'reasonably practicable' of contemporary legislation, something that the Victorian mill owners would probably have heartily welcomed.

Although these arguments were often made from the perspectives of those profiting from labour, the workforce were also brought into the production debates of the early legislation, positioned by some on the same side as their employers. For example, the argument was made that people wanted to work for as long as they could (irrespective of safety concerns), simply so they could earn as much money as possible within their working day. As Sir J. Graham (1844) stated:

> … that adult labourers, stimulated by the honest desire of earning as much as they are able, because wages are generally paid by piece-work – so far from going to those masters where time is limited, invariably prefer the establishments where most work is done.

This shifts the workforce onto the side of production, bringing further support to the valuing of industry and output above all else, whilst also negating concerns of long working hours, working conditions and indeed safety. However, this argument does not itself get to the core of this issue – that of piecework, which is simply assumed common practice and readily accepted within hazardous industrial workplaces, despite the obvious rewards to the owners at the expense of the workforce.

This situation is arguably even more valid for today's workforce; construction trades are *still* often paid on price or measure, and therefore relationships between production and safety remain problematic. However, piecework has not always been adopted so readily by industrial workforces. For example much has been written about the hostility to piecework in Germany, where it was traditionally seen as an upset to the social contract between the worker and his manager, and attempts to introduce it in the 1920s resulted in the development of the slogan *Akkord ist Mord* – piece-work is murder (Geerling and Magee 2012) – clearly acknowledging the close relationships between price, speed and safety. Yet as discussed in Chapter 2, in the UK piece or pricework remains a core payment structure on sites, particularly as work cascades down our often lengthy supply chains.

The balance between safety and production, and the role of government and legislation in its control, was neatly summed up in these early debates by Sir G. Grey (1844):

> Legislative interference … how far it could be carried consistently with safety – and how far for the interests of the operatives on the one hand, and the interests of the country on the other …

It is somewhat ironic that the term 'safety' actually appears within this text, but it is the safety of the country, the economy and the role of parliament, and not the worker to which it is referring. And indeed from these early debates onwards, judgements of time and money have become a central part of the lexicon of safety; safety versus production is something that has accompanied safety legislation from the very start of its existence. And it now exists explicitly within the contemporary concept of 'reasonably practicable' found within the Health and Safety at Work etc. Act 1974. Although safety standards in the UK have undoubtedly improved from the times of the industrial revolution, the positioning of safety versus production can still be found throughout contemporary construction operations, readily revealed by the way we position and talk about safety and work on site.

Production in Practice

The segregation of safety from work practice and productivity enables the development of relationships between them that can be traced back to these earliest of arguments; safety readily becomes involved in confrontational and negative relationships with the processes of production.

The case 'against' safety is often based on this type of argument:

> ...it stops the job half the time, it gets too carried away.

The segregated entity of safety has no place within practice, and for this operative it even has the power to stop production. Safety has become the practice of just safety itself, and is firmly and clearly positioned as a hindrance to his work. The workforce is on site to work – and to work efficiently – not to work under the restrictions that safety can place on construction tasks, be they work methods designed to reduce risks, PPE to protect, or planning measures that require preparation in the form of records, discussions and signatures. This reveals one of the fundamental shared understandings of the construction industry itself, as suggested by the considerations of the site context made in Chapter 2, that of the fundamental 'truths' of construction site context is quite simply that *production* is king.

Although safety practices such as risk assessments are arguably not too onerous when considered practically within the scope of site work and the potential dangers found there, they are still not production. As found in Chapter 5, risk is often devalued within the shared understandings of safety, and is often accorded the status of a considerable hindrance when positioned within the production-driven reality:

> The hassle of safety, sometimes it outweighs the actual job…

Evaluations are made of this relationship between the practice of work and the practice of safety, and value judgements not dissimilar to those the Victorian mill owners made between outputs and workforce welfare. Within this comment above, efforts to 'practice' safety are even positioned as more onerous than the work itself, and its negative role alongside that of production is clearly established. This shared understanding of safety will resonate readily for those involved in safety management at the site level. For example, the time it takes to follow safe working procedures is often used to justify violations, and as a site manager, the number of times I heard:

> Just don't look for one minute…

from an operative standing or reaching or balancing on something they shouldn't, and from which they could easily have slipped and fallen, was certainly a daily occurrence. Safety is positioned as the enemy of production, and where time is money, especially for those paid on pricework, a minute compared to the half hour it might take to find the correct access equipment is certainly worthy of consideration.

Yet what is often missing within such evaluations, and considerations of safety and production, is the understandings of any *good* safety may be doing within the workplace. Potential consequences in terms of accidents or incidents that might occur if safety were not in place are also relatively rare within this particular construct of safety; indeed potential repercussions of safety violations do not seem to even exist within this context. That so many people suffered to secure safety legislation in the UK clearly has a shorter shadow than the concerns for productivity and economic gain that fought against it. And it could be suggested that safety is still not winning this battle of production in practice.

And this is further compounded (as all battles throughout history have been, and indeed still are) in that it is in the majority being developed and perpetuated by the winners. This understanding of safety remains the luxury of those whose violations of safety for the gains of production were successful. The problems of safety as a non-event again become realised. Those who took a risk that did not result in an accident are readily able to tell the tale – those who didn't, can't. This therefore becomes the dominant discourse, and one that will tip to the balance to the side of production and not safety.

Site management are only all too aware of the ongoing challenges between safety and production – as they too also work and enact this reality every day. Even more complexly, they have to manage safety within the parameters of productivity, and so their understandings of safety within this context are all the more conflicted. Within management practices where safety is prioritised, explicit considerations of this relationship can therefore be found, for example induction slides often boldly state that:

Having a safe place of work is more important than production

Yet despite such direct challenges to the priority and dominance of production in its associations with safety, this text actually positions safety within a reality where this prioritisation is *not* inevitable. Site management are all too familiar with the challenge of safety versus work, as evidenced by this induction slide, and are drawing on this shared understanding to seek change and the prioritisation of safety. That for this slide the focus was on a 'safe place of work', rather than 'safe work' could also reveal a resignation to seek to construct change around what *can* be changed within management understandings of safety. The physical work environment in terms of place and space may be more easily managed than the much more complex challenges of changing a shared understanding of safety within the workforce themselves – and reprioritising safety over production within their work practices. Such a shift may simply be too large to attempt within a site induction, indeed there would need to be an acceptance and subsequent enactment of this reality by those within the workplace for it to emerge and perpetuate as a new 'truth' of the construction site environment.

Yet at times our existing site environments, and the work practices within, are themselves positioned as negative influences on the practice of safety. Work becomes the barrier to safety, rather than safety to work:

It all comes down to earning money, doesn't it?

Common site management processes are often constructed negatively alongside any positive implementations of safety in practice; either safety must be sacrificed for production or production sacrificed for safety.

The use of subcontracting, sub-subcontracting and sub-sub-subcontracting (as noted in Chapter 2) creates supply chains along which production pressures flow, and ever increase as another slice of profit it taken from each tender price the lower it goes:

> …the subcontractor nature, all he's interested in is earning money

This 'nature' is used as justification for their behaviour within the site environment and poor engagement with safety practices. Yet despite the subcontractor often being seen as a different 'animal' to the rest of the site team and to any directly employed main contractor's operatives, this is not without understanding, and subcontractors are often considered sympathetically. The site reality is one where some are under more production pressures than others, and this is understood, and even becomes accepted – this hierarchical nature of safety in explored further in Chapter 7. That the participation of subcontractors in safety in practice is inherently bound up with money, or rather the traditional payment structures and processes within the site context, often positions the work itself as the dominant challenge to safety.

From the subcontractors' perspectives, this understanding of safety readily emerges:

> Everything's got to be a cheap price, so if you can get away with doing something slightly unsafe, but it's done quicker, you can make a little bit of money out of it.

Again the common practices around payment for production within the site context are positioned in battle with safety, and used as justification for violations – which naturally have been accorded minimal importance and any lack of consequence. This reality is one where taking a few small risks could mean benefiting employers in terms of speed and profit. Yet again, production is king.

Whether the main contractors' or the subcontractors' perspective is used, there is a highly consistent 'truth' emerging here about the construction site reality; it is a reality where the common management practices of lowest cost tendering, subcontracting, payment on price and the need for production, are all drawn into the social constructions of safety on sites, and unsurprisingly have a negative influence on safety practice. These influences can be seen from both sides of the construction 'site coin' and the alternative versions of the same environment within which both parties interact. Whilst management seek to re-position safety as the:

Number one priority!

This familiar slogan is immediately challenged by shared understandings of how safety *really* works within the construction site environment. And this battle of safety versus work and work versus safety is a circular one; either the practice of safety is positioned as a negative influence on production, or the practices of production as a negative influence on safety. Whether a balance can or even should be sought is up for debate – but the ongoing tensions between productivity and safety remain a recognised aspect of construction site life, and their ready manifestations within the way safety is talked about and understood in terms of practice, serve to further highlight its considerable scope of influence. Much more significant changes need to be made in terms of higher level management practices and work structuring if safety is ever to truly become the number one priority.

Summary

When safety is considered through management systems or practices, it often emerges as an entity – something that can stand alone and exist and be managed apart from work. On sites, this leads to understandings of both segregation and engagement; safety can be set apart from work practices, but also embedded within them. This contradictory understanding of safety is to be expected, and indeed reflective of the 'simple complexities' of everyday life and how we understand and develop safety within this context.

Yet the oldest arguments around safety and production, that outputs would suffer and bring economic harm to organisations and the country as a whole, that were set against bringing in *any* form of industrial regulation to benefit the workforce can still be found on sites today. It would seem depressingly little has changed. In constructing safety as an entity, it is still positioned as a challenge to work, something secondary to the prioritisation of production, and more importantly the close associations with time and money. However, that safety is itself hindered by common construction industry work practices is also worthy of note – reflections from the workforce suggesting that safety is indeed a valued, albeit struggling, part of everyday site practice.

This understanding is further reinforced by the common practice of bundling up: safety, health, environmental, quality – all squashed together under the remit of one poor manager or advisor, and all forming part of the construction industry's non-productive outputs, amongst which safety still struggles to become the number one priority.

The inconsistency and incoherence of these realities, and the subsequent variations in the positioning of safety within work practice are of concern – and go some way to explain the presence of disassociation and disaffection amongst the workforce in terms of safety on sites. If there is inconstancy in the relative values placed on production or safety, on its management in practice or simply how safety is labelled on sites, it should also be unsurprising that engagement within the workforce is limited. Indeed, when considered alongside the common construction of safety as an entity, there are significant repercussions in terms of ownership and responsibility, and as a result current safety management has had to rely on both engagement *and* enforcement to ensure compliance.

References

Geerling, W. and Magee, G.B. (2012) *The Introduction of Piecework in East Germany 1945–1951*. Department of Economics Discussion Paper 07/12, Monash University.

Graham, J. (1844) House of Commons Debate 15 March 1844 vol. 73 cc1073-155, Hansard Online.

Grey, G. (1844) House of Commons Debate 18 March 1844 vol. 73 cc1173-267, Hansard Online.

Health and Safety Executive (2007) *Managing Health and Safety in Construction – Construction (Design and Management) Regulations 2007 (CDM) Approved Code of Practice*. HSE Books, Suffolk.

Health and Safety Executive (2014) *Health and Safety in Construction in Great Britain* [Online]. Available: http://www.hse.gov.uk/statistics/industry/construction/construction.pdf [25 September 2014].

Hyatt (1833) House of Commons Debate 05 July 1833 Vol. 19 cc 219-54, Hansard Online.

Lingard, H. and Rowlinson, S. (eds) (2005) *Occupational Health and Safety in Construction Project Management*. Spon Press, London.

Putson, D. (2013) *Safe at Work? Ramazzini versus the Attack on Health and Safety*. Spokesman Books, Nottingham.

Rawlinson, F. and Farrell, P. (2010) UK construction industry site health and safety management: an examination of promotional web material as an indicator of current direction. *Construction Innovation: Information, Process, Management*, **10**(3), 435–46.

Chapter Seven
Engagement and Enforcement

Getting the man and management into the game and working for safety because they wanted to rather than because they had to.

Palmer 1926: 9

The way the construction industry and its sites are organised inevitably leads to a hierarchy of management; various supply chains of subcontractors, their subcontractors, and their subcontractors' subcontractors all feed upwards into the principal contractor or project management organisation taking charge of the site in that role. Principal contractors therefore provide both a corporate management approach and management through the project site team, including some level of on-site supervision. Below this level, the supply-chain subcontractors provide management for their own work packages, as well as on-site supervisors or foremen who in turn manage the operatives, the ones who actually undertake the work itself.

However, this hierarchical arrangement is not always beneficial for safety management (Lingard and Rowlinson 2005), and it can create disharmony within sites, as conflicts in terms of individual goals and production patterns are developed through the way our work is segregated. Subcontractors, with their focus on a particular work package, can be keen to complete their work and their work only, with little regard for preceding or following trades. Long subcontractor supply chains can also result in price squeezes that cascade downwards, and consequently time and money become ever

Unpacking Construction Site Safety, First Edition. Dr Fred Sherratt.
© 2016 John & Wiley Sons, Ltd. Published 2016 by John & Wiley Sons, Ltd.

more magnified and critical to the realisation of any profit the lower you go. For some on site, working on very tight profit margins, coordination and cooperation can become more difficult to accommodate, and the prioritisation of work versus safety can become all too quickly manifest.

Nevertheless, principal contractors *must* strive to maintain safety throughout their sites, despite this problematic context and hierarchical 'structure' of safety on sites.

In order to achieve this, safety management on large sites has developed an identifiable 'two-pronged' approach; one which seeks to enforce the rules and regulations of safety as required by law and safety management systems, and one which seeks to engage workers throughout the supply chains. Safety engagement encourages participation and involvement in the practices and management of safety on sites, something akin to the concept of a 'safety culture'. Ideas of safety engagement are not particularly new – the quote that introduces this chapter is readily identifiable as a quest for safety engagement, and this comes from a paper published in the USA in 1926 on the *History of the Safety Movement* as it was then – but the construction industry, certainly in the UK at least, has only relatively recently sought to actively promote engagement with safety; safety slogans, targets and branding providing the prominent outward face of the safety management programmes that now form an integral part of safety management on sites.

However, engagement and enforcement are not mutually exclusive and, when unpacked, it can be suggested that through these two different 'voices' of safety, different realities are developed and again incoherence and inconsistency of safety can be found; sometimes rules are made to be broken, punishment is a necessary evil, whilst engagement can prove problematic in its impact, raising far more complex issues of the ownership and responsibility for safety on sites.

Engaging with Safety

Engagement of the workforce is actually set out in the safety legislation of the UK. As Clause 2(6) of the Health and Safety at Work etc. Act 1974 states:

> Cl 2 (6) It shall be the duty of every employer to consult any such representatives … to cooperate effectively in promoting and developing

measures to ensure the health and safety at work of the employees…

Although engagement is not the term used here, it is a recognisable part of an active and dynamic process of safety. Further clarification of the need for worker engagement has been included within the Construction (Design and Management) Regulations, the 2015 revision of which explicitly makes reference to engagement as part of the principal contractor's duties:

Principal contractor's duties to consult and engage with workers
14. The principal contractor must—

(a) make and maintain arrangements which will enable the principal contractor and workers engaged in construction work to cooperate effectively in developing, promoting and checking the effectiveness of measures to ensure the health, safety and welfare of the workers;

Again, safety here is the outcome of an active process, with engagement positioned as the 'cooperation' needed for aspects of its creation. Safety engagement has therefore become part of the management process to ensure safety within the workforce. But whilst engagement is a very easy term to use, safety committees can be readily established, often enhanced with the lure of chocolate biscuits and tea to encourage attendance, the actual creation and enactment of 'engagement' out on the site as part of an everyday reality can be much more complex.

A prominent way engagement has become manifest on construction sites draws on its associations with participation and cooperation, the social aspects of safety, which in turn are closely associated with the now established idea of a 'safety culture'. 'Safety culture', in various guises, has become a key safety management tool amongst large construction contractors (Biggs *et al.* 2005; Ridley and Channing 2008; Dingsdag *et al.* 2008), seen as the natural 'next step' in safety management development after the implementation of a safety management system within an organisation (Hudson 2007; Meldrum *et al.* 2009). The Health and Safety Executive also actively encourage a proactive 'safety culture' on construction sites, viewing its development as essential to improve the safety record of the industry (Health and Safety Executive 2000).

But culture is of course a highly problematic term (arguably one of the most complex concepts in the English language!), even more so than safety itself. When we start to explore and unpack what we understand culture to be and indeed how it works in our societies, myriad problems, philosophical positions and challenges emerge. However, that does not seem to have prevented the construction industry on a global scale from adopting 'safety culture' as a prominent weapon in the safety management arsenal, and any problems of understanding have certainly not stopped the enthusiastic quest on construction sites for a 'positive safety culture'.

The original concept of an occupational 'safety culture' had close links with the ideas of cause and effect around accidents, where unsafe behaviours were often cited as evidence of a 'poor safety culture'. Indeed it has been argued that the popularity of 'safety culture' as a concept was in part due to the ease with which accident investigations could so readily conclude that poor or inadequate 'safety culture' was the underlying cause (Baram and Schoebel 2007) – again a box could be ticked on the incident investigation form and we could all move on. However, agreement as to what a 'safety culture' actually is, how to measure it, or how to effectively develop one has not yet been established within the construction industry. There are a large variety of definitions, models and processes which attempt to answer these questions (some examples can be found in the work by Mohamed 2002; Baram and Schoebel 2007; Hudson 2007; Ridley and Channing 2008; Hartley and Cheyne 2009; Maloney 2011; Wamuziri 2011 to name a few), but as with the ideas of 'culture' in many other fields, the debate continues.

A common way of thinking about safety culture is by considering and implementing the accepted ways of establishing one within your organisation or on your site – and unsurprisingly this is the approach most commonly taken in practice within the construction industry. Generally, safety culture management seeks to go beyond the implementation of legislative requirements, operative and management training and formal safety management systems, and draw on other aspects for success. Wamuziri (2011) identified four key factors as prominent components of a positive safety culture specifically within the construction industry context:

Firstly, top down senior management commitment is commonly cited as an essential ingredient in the establishment of a positive safety culture on sites (Lingard and Rowlinson 2005), and by some as the most important factor affecting safety culture (Hughes and Ferrett 2007). Commitment is required not only from the main contractors, but also throughout their supply chains. In practice, the main contractor's corporate 'voice' has permeated safety management in the form of 'safety leadership', speaking to the workforce through safety posters and training.

Secondly, communication and involvement of and with the workforce in the practical methods of safety management on sites is sought. This can involve several different site practices, including Daily Activity Briefings held every morning to discuss and plan the day's work including the key safety risks, the appointment of worker safety representatives, the use of anonymous safety suggestion boxes for safety improvements or near-miss reporting, and safety committee meetings between workers and the site management.

The third factor involves the creation of a 'no-blame culture' (Mohamed 2002; Dekker 2007) on site. Assurances are given that workers will not be blamed if they are involved in an accident, an approach designed to encourage accident and near-miss reporting and enable open and honest communication around safety (Gadd and Collins 2002) between the workforce and management. This approach is based on a move from 'human error' as the causal factor in accidents to a systems-based approach, which would enable underlying reasons and causes to be established beyond the incident itself (Health and Safety Executive 2005). Such an approach is a paradigm shift from the traditional retribution, punishment and reprimand that is often associated with an accident or near miss on sites.

The fourth factor involves the implementation of 'safety management programmes', which seek to bring together all of the above, to disseminate organisational policies and practices, and are used to define and articulate the safety culture change in process. The use of training and branded 'safety propaganda' are used to win the 'hearts and minds' of the workforce, and often speak to the workforce on a personal level, asking them to take responsibility for their own safety; encouraging the desire to choose to work safely, rather than compelling safe working by enforcement and policing. An example of such propaganda can be seen in Figure 7.1, the slogan on this high-viz vest personalising safety and seeking to directly link it to the worker's actions. The approach is not reliant on rules or paperwork, but rather respect and expectations, and is based on effective communication, worker education and engagement and creating an environment which can challenge the way work is undertaken on sites (Worthington 2007).

However, these programmes have not been without their critics. There was a vocal backlash in particular to the behavioural elements of early manifestations of the safety programmes, both in the USA and the UK, with claims that they tended to 'blame the worker'

Figure 7.1 Think Safety Act Safely

rather than focus on potential hazards and unsafe conditions within the site environment (Frederick and Lessin 2000). Indeed, there is little firm evidence of the success of these types of programme, despite positive reports about implementation on large sites (Health and Safety Executive 2008), and the importation of some programmes direct from the USA may be an optimistic attempt to procure an off-the-shelf solution from a different country, with very different social understandings, including those about safety.

The effectiveness of these 'safety culture management' approaches has been explored far less within academic research than the broader concept of 'safety culture' itself (Biggs et al 2005). A notable exception is the work by Hale *et al.* (2010) which highlighted the practical difficulties of establishing positive or contributory relationships between such 'safety culture' interventions and improvements in practice. Hale *et al.* also concluded that although achieving positive change is hard, both in terms of its actual implementation and its evaluation, workforce engagement, communication and management commitment were emphasised as influential, whilst training and branding were 'at best necessary, but not sufficient requirements for improvement' (Hale *et al.* 2010: 1035). Yet despite such concerns around effectiveness, and a lack of evidence of positive change in practice (Health and Safety Executive 2008), various 'safety culture' focused management programmes have frequently been adopted as a key tool on sites and implementation of these practices and policies are commonly found on large UK construction sites.

Safety Propaganda

Ignoring the negative connotations that have come to be associated with the term propaganda, it is actually a highly appropriate way to describe the methods and approaches of the different safety management programmes currently in use within the construction industry. As Edward Bernays (2005: 53) pointed out, propaganda 'does change our mental pictures of the world', something that resonates closely with the goals of such programmes as well as the ideas of how safety works that have been explored within this book. However, in terms of *effective* propaganda, when the common approaches used for the promotion of safety through engagement are explored and unpacked from a constructionist perspective, some potential challenges arise.

For example, safety propaganda communications are often personalised, adopting the notion that workers engage at a greater level with safety messages if they understand the consequences of poor safety at the personal level (Biggs *et al.* 2005). The sign in Figure 7.2 displays a very commonly used phrase in safety engagement:

Figure 7.2 Make Safety Personal

Yet the slogan in Figure 7.2 is a rather a bold and bare mechanism of engagement; through a direct command it demands a response from its audience in their own understandings of safety on sites. However, despite the function of this sign to link safety to the 'personal', what it actually does is construct a situation where safety is *not* personal right *now* – rather it needs to be made so from this point forwards. This has repercussions for the individual within this relationship – as safety is *not* personal to them right now, it is therefore *not* actually their responsibility either.

This demonstrates one of the more fundamental problems of engagement through the signage and slogans of safety propaganda. Much like those around the prioritisation (and therefore segregation from work practices) of safety, the isolation of engagement as an explicit process often inherently disengages the recipient from being or even becoming part of that process.

There is also the potential that safety propaganda remains just that – slogans and throwaway phrases with little impact or instigation of change, unable to engage with existing shared understandings of safety and the realities in which they operate. By necessarily reinforcing an understanding in which safety is *not* personal, where it is something abstracted from work, to be managed by someone else further up the site hierarchy, the social influences of such slogans may be superficial at best.

Another way engagement is often sought around safety is through the repeated use of personalised pronouns, as found in these examples taken from various site inductions:

Our commitment to your safety on this site …

You have a voice and we will listen!

Your input and cooperation is vital in worker and workplace safety

Whether used in signage, inductions or other safety literature on sites, the use of 'You' and 'Your' is commonplace – another, more indirect, way of making safety personal. However, in doing so the limitations of language again come to the fore, and this personalisation also inevitably constructs a segregated position; the worker taking the role of 'you' and 'your', and the main contractor the role of 'our' and 'we', as the voice of instruction and management. In seeking engagement the opposite has been achieved; through the linguistic constructions necessary to develop personal and individual messages within the texts, inherent segregation becomes the unavoidable consequence.

Although this is not something that has gone totally unrecognised; attempts are often made to contextualise personalisation and to try to overcome the segregation/engagement paradox. For example as noted above, a shared 'workplace' is made as the contributory goal of the engagement process rather than any individualistic gains, and workforce committees – although a rather mechanistic manifestation of engagement – are often positioned as a bridge between workforce and management. In some cases redress is sought through explicated awareness of that segregation, as found in the second example above, whilst it is recognised that there are two separate parties involved, the communication channel between 'you' and 'we' is clearly highlighted and emphasised as part of the safety hierarchy of the site.

It is an inevitable consequence of the approaches of participation and engagement that the alternative is also constructed. The reality *before* such engagement approaches existed becomes one where workforce participation in safety was considered irrelevant and the workforce were there simply for productivity, rather than stopping the work for every little safety concern. This closely links to existing understandings of safety versus work and work versus safety, and to some extent maintains the past, where work was most certainly prioritised, with this new present. Yet it is arguable that this reality still has to be constructed in order to ultimately purge it, and to develop change within the dominant shared understandings of construction site life.

The way safety propaganda is presented and delivered to the workforce can also be problematic. Distinct from training or competence, safety engagement is more often sought through an approach grounded in education – which is often, again somewhat inevitably, disassociated from practice.

Those who have experienced safety education programmes in their everyday working lives often make both positive and negative associations, interconnected by a highly consistent way of talking about them; an initial construction of the positive aspects of the programmes juxtaposed, often within the same sentence, with a negative qualification – the 'but …' of safety education. Initial praise for the programmes ensures the speaker's own self-alignment with what could be considered the 'accepted social norm'; the aim of such programmes is to reduce accidents on sites and so it would be very hard to challenge such a philosophy directly. However, once initial positive positioning of safety education has been made, negative qualifications are then more free to follow, as this operative said:

> Just because you've seen a video and had a bit of a chat … I don't think it necessarily instructs you to be safety conscious.

Here personal understandings of safety have been positioned alongside the programme delivery, this operative has found it lacking; site safety is not something that can be readily associated with and enhanced by such 'educational' methods, similar to the way safety training in a non-practical environment is also readily dismissed as irrelevant. The method of delivery is incoherent with the way this operative understands how safety works, and in his reality the two just don't match up.

Audience responses to these safety management programmes – the worker above being a prime example – are of critical importance to their impact, as this site manager said:

> The right people get it … but the normal person, they just want to get their money, do the job and get away…

Here, a specific site reality exists with two different types of players; the 'right' people and the 'normal' people. In the realm of the 'right' people, the programmes work and success can be achieved. But 'normal' people, carrying out their construction work for their weekly wage, operate in an alternative reality and are not interested in safety programmes or training. The labels the speaker uses clearly indicate who they find more frequently on their sites, and what they consider 'normal' within their understandings of everyday construction. But this does not present a welcoming context for any safety education programme, and suggests a challenge for those looking to deliver them within this 'normal' site reality.

Safety education is also something that is often understood to be fluid – a version of safety that can be seen as a 'magic potion', which unfortunately will eventually wear off:

> It helps, but the further you go from doing the actual course, the less and less you stick to it…

Again, a positive appreciation of the programme is followed by a 'but …', however, this but is different to the normal challenges of work practice or the reality of site. Rather here, the negative construction of the safety programme has been developed through self-reflection, in which safety education is constructed as a treatment which becomes less influential the further away in time you travel from its initial application. Emphasis is placed on the active participation within the safety course, rather than long-term development of practice, and the education gained there is seen almost as something that fades over time – as it doesn't fit with the everyday experiences of site life it starts to tarnish and rub it off as work life goes on.

The idea of safety 'wearing off' can also be found in the safety propaganda itself, for example induction booklets often ask the workforce to:

Please remember … safety is important

suggesting safety is something that quite easily can be forgotten about. Yet this statement was included within a safety management system safety guide, something that would actively seek to position safety as central and embed it firmly within work practices, making it rather out of place. Alternatively, this can be interpreted as an insightful awareness of the reality of safety on sites in which this particular document is operating. The everyday site environment, and the realities of work versus safety, have been accepted here, and the fact that safety 'wears off' has been acknowledged as a somewhat inevitable consequence.

Safety propaganda can be found on most large UK construction sites in some form or other, the programme branding on posters, hard-hat stickers, high-viz vests and hoardings seeking to engage the workforce in safety. Yet despite the terms of engagement – the we and us – seeking to harmonise and unify, they also inevitably segregate, reinforcing the distinct management hierarchies of the site. That the delivery methods of the safety propaganda seek to step away from the norm and into a more educational approach can also conflict with the way the construction site likes to learn; the lack of association with practice impacting their effectiveness in a similar way to classroom-based safety training for the more practical aspects of construction activities (see Chapter 4). The idea that safety propaganda can also wear off is highly relevant – faded posters or a one-off induction or training session are not understood to be long lasting, clearly demonstrating the potential for a lack of engagement with the process.

Yet engagement remains a prominent way safety is managed on sites, whether through bespoke and branded safety programmes, through various approaches made under the umbrella of 'safety culture', or simply grounded in the legislative requirements for workforce committees and feedback boxes on sites. However, this forms only one half of the safety management approaches commonly found on large sites. Safety engagement also exists in a reality alongside that of safety enforcement, where the need for the site hierarchy – the them and us – to impose the safety rules born of legislation and basic operational practices becomes a far more prominent and influential aspect of safety management in practice.

Enforcing Safety

Enforcing safety is all about the rules. Safety rules often form an inherent part of any safety management system, they are read out during site inductions, referred to in method statements and prominently displayed on site notice boards. Rules are often grounded in legislation; in the UK safety management, by necessity, sets rules, practices and processes at the very minimum to meet the legislative requirements (Howarth and Watson 2009). For example, the need for risk assessments and their communication before work starts is a requirement of many different pieces of legislation, as is the need to reiterate the requirements of the Health and Safety at Work etc. Act 1974 Clause 8:

> No person shall intentionally or recklessly interfere with or misuse anything provided in the interests of health, safety or welfare...

Rules are a way of reminding site workers of the law; the incorporation of the language of legalese within the rules, taken from the safety regulations themselves, gives them the voice of the 'ultimate' level of management and ascribes the rules a higher authority within safety management than that of merely the site itself. This direct translation to the site environment is unsurprising given the volume of highly specific legislation applicable to construction sites, and the way this legislation is often used within safety management systems and site rules ensures nothing is 'lost in translation'.

Site rules can take a variety of forms, from a typed A4 list of 'dos' and 'don'ts' to much more sophisticated printed booklets, however, what the rules *are* is not necessarily important – the content of site rules does not usually vary much from site to site (mandatory PPE to be worn at all times on the site, dedicated walking routes, the tickets required for the use of plant and machinery, the need for temporary works inspections and scaffold tags etc.) – what is far more revealing is *how* they are positioned with regards to safety, and what repercussions this has for how safety works on sites.

Many large contractors have distilled their site rules into professionally printed booklets to be distributed on induction, supported by posters to be put up around site. These documentary artefacts of safety are produced within a corporate rather than a site context, and are often positioned as a part of wider organisational safety management systems and safety management programmes. Such site safety guides often present the 'rules' through the dichotomy of:

✓ Always   ✗ Never

partnered to establish clearly illustrated parameters for behaviour. Alternatively rules can be embedded within specific site hazards, with frequent reference to a 'safe system of work', directly associated with practice and positioned as descriptive of the processes of work. Although perhaps not directly expressed as such, rules are found in the advice of 'do' or 'do not' which again create the requirements for behaviour and action.

Yet whilst such corporate documentation is willing to clearly set out what is 'right' or 'wrong' with regards to safe behaviours, what it does not acknowledge is that rules and regulations are also part of a much wider social paradigm which advocates compliance and rule following – but which also incorporates enforcement, potential violations and consequential punishment. Site rules must be contextualised within this wider concept of the legal framework of governance in which UK society operates, and what is understood to be the way rules work in practice.

However, reference to enforcement or punishment is almost always found to be lacking within the voice of corporate safety management. Despite the presence of prohibitions which clearly establish benchmarks for violation, further clarification of what enforcement is in place or what punishment will be meted out for lack of compliance is often totally missing. Instead, this corporate voice operates in a reality where there is no need for punishment as an aspect of safety enforcements; rules and prohibitions are made with no recompense for potential violations, because in this reality people simply don't break the safety rules. This can be linked to a key aspect of the safety management programmes, which position violations that result in safety incidents within a 'no-blame' culture; and as a consequence there is no punishment to be meted out. Through the corporate voice, this lack of punishment can even be extrapolated to a preceding stage; corporate management constructing a reality that does not accept or even acknowledge any violation that could lead to blame.

Whilst this could of course be a conscious and deliberate approach made by the authors – if the documentation is assumed to be operating in a reality where violations therefore do not occur, any form of punishment becomes unnecessary – the way site rules are positioned and managed by those working on sites would suggest that this is simply not the case. Indeed, site-produced documents, posters and inductions with reference to the site rules certainly include and in fact more often than not prioritise a:

Disciplinary Code; Breach of H&S rules on site e.g. misuse or lack of PPE

The prominent inclusion of violation, discipline and punishment associated with safety within site-produced documentation reflects something of the reality in which they are operating. This construction site is a place where violations do occur, and therefore necessitates the need for a 'Disciplinary Code'. Often highly detailed in terms of the disciplinary practice and process associated with enforcement of the rules, punishment is not only directed at individual violators, but compliance also becomes the responsibility of the organisation, and the violator's supervisors, managers and directors are also frequently included in the threats of disciplinary action. This suggests development through practice; to discipline the individual has historically proved ineffective and as such construction site punishment requires more convoluted and detailed processes incorporating the violator's organisation and their employers.

Examples of these two different 'voices' in the enforcement of safety can be readily found within site signage – in the differences between the corporate, professionally produced, safety propaganda, and the signage produced in the site offices on the printer and laminator, or even handwritten.

Safety propaganda often seeks to engage with safety, even when espousing site rules, for example this corporate safety programme sign asks that its audience:

> Choose ... to work safely ... not to enter segregated areas ... not to jump barriers

In making these statements, this sign is operating within a reality where people did perform these behaviours, do act unsafely and specifically could violate the access provisions. So to counter this, the sign firstly employs ideas of engagement with safety, and seeks to develop safe practice through appeals to the individual's autonomy, although this is juxtaposed with the contradictory approach of safety as enforcement – the prohibitive 'not to' rules also located within the text. However, despite the acknowledgement of violation, there is no acknowledgement of punishment or retribution should the operative choose not to follow this guidance, instead the sign simply encourages through the voice of engagement.

This can be compared to a sign produced in the site offices, laminated and cable-tied to the walkway barriers, addressing one of the same safety violations, and states:

> Barriers and yellow walkways are there for your safety and protection and should not be moved ... anyone found to be moving barriers or walkways will be subject to disciplinary procedures!

This sign is also operating in a reality where people had violated the same access provisions, but goes about it in a very different way. This sign is very much bound up in addressing a specific previous action by others and establishing future control. The need to construct and display such signs clearly suggests past non-compliance with this particular site rule, and the threat of 'disciplinary procedures!' implies the need to reinforce compliance and so seeks to address the audience in straightforward terms of punishment avoidance. This sign is involved in the enforcement of safety, and draws on the rules, violation *and* punishment to undertake its safety management function.

When the two voices are compared, the site voice appears harsh in its approach, issuing threats as it constructs safety as enforcement, however, this is clearly born of frustration and a very real problem in practice – in fact one that is itself acknowledged in the corporate safety propaganda. Although both signs are operating within a reality where violation of safety rules occurs, it is only at the site level where this is made manifest and with it comes the need for discipline and punishment. Indeed, throughout the documents and signs of the site management, it is common to find protracted and highly detailed dialogues around the enforcement of safety, detailing the punishment mechanisms for violation of certain safety rules and directly addressing the workforce to threaten them into compliance.

The distinctions between these two voices create a dissonance in the way the enforcement of safety works. Whilst those at the higher corporate level seek to develop and position safety only positively, through no-blame cultures and realities intolerant of violation to the point of denial, those who manage and partici-pate in construction site practices on a daily basis at site level instead have a version of safety firmly positioned within a reality of rules, violations, enforcements and punishments. Yet this latter approach also has the potential to create an understanding, or rather misunderstanding, that safety *is* the rules, rather than any wider considerations of safety and practice. In fact, when the safety rules are explored in more detail, their associations with safety become rather irrelevant and the enforcement of safety is much more bound up in issues of discipline and punishment on a societal level, rather than the potential consequences of any safety violations themselves.

Rules Made to be Broken?

At the site level, the safety rules become something all too readily associated with their violation – rules that are made to be broken. For anyone with any site experience, safety management and the need to enforce the safety rules is something that happens all the time. From access equipment issues (handrails disappearing from ally towers, the wrong ladder in the wrong place, or just reaching a little too far on those podium steps), to working outside of risk assessments because something has (inevitably!) changed, to the banality of PPE violations (missing glasses or gloves or the lack of a hard hat because someone was working in a really tight space), safety management, however minor, often becomes an integral part of every site walkround.

With safety violations such a common occurrence, it is not surprising that the punishment and discipline that goes along with them is also prominently associated with safety, as this foreman notes:

> It's more to do with whoever's running the site .... if they're giving out red and yellow cards, people see them doing that, it makes them more aware that people are coming down hard on health and safety.

Because safety is something that is inevitably violated, it is the site management's subsequent actions that become important in how safety works – and here punishment is all important in how safety works on this particular site. Rather than looking to management practices that might support or encouraged safe actions on site, safety has become something to be controlled and created through discipline, based on a football-style card-based punishment system. Safety is something that can be come down 'hard on' – the language used also closely associated with discipline and punishment mechanisms used for enforcement.

This creates and perpetuates an understanding of safety as the safety rules, which are in turn inevitably broken and therefore closely linked to punishment. This naturally embeds safety within the site hierarchy, through the allocation of 'safety' roles for managers, supervisors and foremen. As the previous quote quite rightly notes, it is whoever is *running* the site that seemingly takes responsibility for enforcing the safety rules to the bitter end. In contrast, as this subcontractor's foreman said:

> ... they might break the odd rule, so I tend to go out and just enforce them a bit more ...

This foreman was talking about his own site operatives and his own management of safety, however, the way he talks about site safety rules gives them a very minor status – just an 'odd rule'. Reference to violations as 'odd' reduces their impact in both frequency and severity, and positions them within a reality where safety violations (which in practice could be very serious in terms of consequences) are frequently minimised through relatively casual talk and linguistic associations. Furthermore, this foreman's enforcement process does not fit within any wider management practice or process framework and no punishment is included within this level of interaction. This subcontractor's foreman is happy to simply enforce safety without further recourse or punitive action to his own site operatives – and ideas of safety as simply enforcement becomes the means to its own end, safety once again simply becomes the safety rules themselves – and at this lower level in the site hierarchy, punishment for any violation has seemingly disappeared.

And as a consequence of this, violations themselves become a key way in which safety works on sites. Within the site environment, compliance with the rules is not as commonplace as the wider social understandings of rules and regulations would suggest; instead the violation of safety rules is simply an inherent and accepted aspect of construction site life, as simply talking to those who work on sites will quickly reveal. Consequently, when we talk about safety, it's not just the rules, but the inevitably violation, subsequent management and enforcement of those rules that come to the fore. Violations are simply a matter of course, such an everyday occurrence around safety that they are often belittled, the way they are talked about reducing their importance and impact in practice. This language of unsafety contains within it an emphasis of the mitigation of any violation; as previously explored in Chapter 5, the majority of safety violations are only 'slightly unsafe', with the 'odd' factor of concern as most behaviour is 'quite safe', which position any violations in negligible terms and with minimal safety consequences. This shared understanding of rules and enforcement does not serve to reinforce the prioritisation of safety within the site context; instead safety becomes minimised, devalued and even negated.

Indeed, safety violations are more often than not seen as bending rather than breaking the rules, with little association with danger or the potential for incident or injury. In fact, when people talk about violations and breaking safety rules, it isn't the potential of an incident or injury that becomes the consequence, rather it is the potential of being caught and punished that is of the greatest concern.

And interestingly, at the site level, alongside the acceptance of safety violation as the natural state of affairs, punishment or some other form of redress for such violations is actually found to be expected. Although violations are often constructed without consequences of personal injury or accidents, they are constructed within a context where punishment might well be the correct course of action, should the perpetrator be caught. The construction site workforce *expects* safety rules to be bent as a matter of course, and if the perpetrators are caught, punishment is certainly due.

For example, this operative was talking about not wearing his gloves, which were part of the mandatory PPE on site:

> I don't mind being bollocked, if it's done decently, and eventually you've got to make sure you do wear them otherwise there's going to be consequences … well you can handle that.

Again violations are expected, rules are bent, and here punishment is accepted as the eventual right course of action. As usual, no association is made to the potential consequences of the violation or indeed any actual need for compliance, why the gloves were needed is not really a consideration and in using the ideas of PPE, it is the gloves that become an associated tool for enforcement, or even safety itself, rather than an artefact of safety in practice. This also enables the speaker to again present the low status accorded safety violations when people talk about them on sites; the violation was minimal therefore the punishment will probably not be too severe. But this speaker also illustrates another common social understanding around rules and punishment on sites, through reference to the acceptability of the interaction with those enforcing the rules, and associated with what could be described as the old-fashioned values of politeness and decency. One of the main concerns could be said to be the need for the violator not to feel violated by the management practices around safety, those of enforcement and punishment, rather than any prioritisation or reflection on their own behaviours which actually broke the site safety rules and why they might be relevant.

When looked at from the perspectives of the hierarchies of safety, the way rules, enforcement and punishment work on sites differs significantly depending on perspective. The formal corporate voice simply sets down the rules, and assumes compliance, with no recourse to enforcement procedures or indeed punishment for violations. In contrast, those managers and supervisors working at the site level closely link rules to their own role in their active enforcement – the violation of safety rules so commonplace that it becomes simply assumed rather than highlighted as anything out of the ordinary, and therefore enforcement becomes simply an inherent aspect of their management role. Indeed, violations are themselves negated in importance, little consideration given to the potential consequences in terms of safety incidents, and in some cases the connection between violations and accidents is in many cases simply never made.

For those on the receiving end of any enforcement, another understanding of safety rules emerges. Although punishment is not necessarily resented by those performing the violation, prioritisation is often given to the social management of the way in which it is delivered. The personal approach of the punisher, and how the punishment was delivered and the enforcement made in practice are all seen as critical to its acceptance. However, this often negates the reasons for the existence of the safety rule itself and avoids all association with the practices that have led to its creation and enforcement, rather the rule appears to be 'made to be broken', particularly in the case of certain elements of PPE which are often drawn upon to give examples of violations in practice.

It can even be suggested that this relatively ready acceptance of punishments for performed safety violations is actually the manifestation of a reality in which the site workforce actually *need* enforcement and punishment in order to positively participate in safety in practice, however much this conflicts with the ideas and understandings of safety engagement. This is an explanation that easily harmonises with the realities that surround the acceptance of violations within the site context, and the workforce, through the bending and breaking of site rules – something made necessary through the wider understandings of safety versus work or indeed work versus **safety** – are themselves seeking discipline and punishment, positioning this as necessary to support and even enable the development of a coherent level of safety management. Safety rules are seemingly made to be broken, and this is something that may actually even be essential to the way safety works on sites.

A Hierarchy of Safety: Responsibility and Ownership

The different levels of management found within the construction site context, and the different ways in which they implement and construct safety on sites create what can be termed a 'hierarchy of safety'. Through the relationships that develop between the violators of the safety rules and those tasked with their enforcement at various levels of management, further considerations of how the responsibility for, and ownership of, safety works in practice can be explored.

Unpacking the enforcement of site safety rules has suggested that both site supervisors and operatives simply accept violations as part of their daily work, and are also resigned to the punishment dealt out if the perpetrator is caught. However, this clearly shifts responsibility for safety up the hierarchical chain – the operatives absolving themselves of any ownership. Rather they are there to work, and if that involves the bending of safety rules, then that is simply something to be accepted. In readily positioning themselves as violators, minimising the potential repercussions of these violations and accepting punishment as it is meted out, the site workforce are able to absolve themselves of any responsibility for their own safety or that of others on sites. They *need* to be punished and managed in order to achieve safe working, reducing themselves to the level of children who need to be controlled and disciplined, yet contradictorily only accepting this control if it is delivered in a 'fair' and respectful fashion.

This then places the site management of the main contractor in an intermediary position; they *must* enforce the site rules, the need for legislative compliance setting certain standards as a minimum, and are often through their organisational policies also seeking safety through both engagement and enforcement. Yet this creates a level of segregation within the site team between the main contractor and the subcontractors, creating 'them-and-us' relationships, and making the construction site a place where enforcement and discipline around safety occurs within an environment structured around two distinct camps; the violators and the enforcers. This is often further reinforced by much of the safety propaganda seeking safety engagement, which also constructs an inherent segregation amongst those working on sites in terms of the main contractor/subcontractor roles needed to seek out and subsequently respond to common engagement approaches in practice.

However, the voice of the main contractor speaking from their corporate office is often very different to that speaking on the site itself. On site, the corporate voice is often in conflict with the reality in which it is operating. Within the site environment, people do break the safety rules and this must be addressed through the discipline and punishment that is often missing from the corporate safety propaganda, grounded as it is in a no-blame environment. However, when discipline and punishment is needed, it is often the threat of this corporate brand that is invoked through reference to safety propaganda, policies and practice, rather than that of the site management team. Instead, the voice of the site management often seeks to disassociate itself for the direct repercussions of responsibility for safety enforcement, for example, a handwritten sign focused on PPE violations noted that:

> … you will be subject to (company name)'s disciplinary procedures

For the site team, this shifts responsibility and ownership of safety up to the level immediately above them in the safety hierarchy, whilst simultaneously developing collaborative associations between themselves and those at the site level; aligning the main contractor's on-site supervisors with the subcontractor's on-site supervisors and operatives. This transfer of control and the setting of the rules to a higher power than on-site management could be symptomatic of the need to maintain a level of harmony within the social environment of the site in order to facilitate the other necessary processes of the sites such as the cooperation needed for quality and timely production. This again generates another 'them-and-us', but at a tangent to the more traditional main contractor/subcontractor divide, and positions it at an on-site supervisory/office level management instead.

However, at the top of this hierarchy of safety, speaking as it does from a position of engagement through safety propaganda, the main contractor's corporate voice of safety seeks to challenge this shift in responsibility, and send it back down the management chain. As this text from an induction slide states that:

> Nothing you do is so important that the time cannot be taken when conducting your works to do it safely

This familiar personalisation creates a direct link between the readers' behaviour and safety. Safety is bound up with practice, the use of 'time' and 'important' positioning it firmly within the reality of productivity (the safety versus work of Chapter 6), yet also challenged the association with reference to safe working practice. However, in positioning the audience as agents of their own actions, it can also be suggested that the responsibility for safety had also been shifted back onto the individual. In establishing the autonomy of the evaluation and decision-making process in 'conducting your works', the individual has also been given the responsibility for any evaluation in terms of safety. This shift in responsibility through engagement and the assignment of autonomy to the site workers can often be found on sites, particularly within safety management programmes and personalised safety propaganda. In urging individuals to 'make safety personal', a level of engagement gives individuals a role of safety in practice, co-opting them into the safety management of the sites and assigning personal responsibilities for safety. These approaches often position the audience of the safety propaganda as a colleague or co-worker in the safety management of the sites, linked to an associated shift in the ownership and responsibility for safety and its management.

Yet this remains in direct contrast to the developments of responsibility identified through the ideas of safety as enforcement. Whilst the workforce have constructed a reality in which they are absolved of all responsibility, to be managed and punished when necessary to deliver 'safety management 'within the sites, safety propaganda directly challenges this through constructions of personalisation, seeking to cascade responsibility and ownership of safety back down onto the workforce. In addition to the incoherence this can create in practice, it is also worth remembering the arguments of Frederick and Lessin (2000), and to remain mindful of the potential for this shift in responsibility to also become a way – either perceived or actually realised – to simply 'blame the workers' for poor safety management. The complexities of responsibility and ownership for safety, as seen from the perspectives of engagement can all too readily become apparent.

But the corporate voice also requires some elements of safety enforcement at times, although this is often dressed in the clothes of safety propaganda, seeking to shift enforcement from the implementation of rules through punishment to the encouragement of individuals to follow the rules of their own volition. For example, a common way of positioning the site rules in corporate site safety guides is to ask that the operatives:

> Choose to follow the safety rules rather than being compelled to follow them

This positions the rules within a context of choice and engagement rather than the traditional rhetoric of compliance and enforcement. Whilst there is still a reality of regulation (the rules still exist as a part of safety on site) which must be adhered to by the addressee, and whether the individual followed the suggested behavioural approach or not, safety has become associated with both engagement and enforcement; the two have become intertwined. Although the rules remain the same, the constructions surrounding them have changed, which in turn leads to the assumed realities of the sites of the safety management programmes as places of an obedient and willingly participative workforce – however unrealistic that may be.

Although safety is being constructed through the language of safety propaganda, directly challenging the 'old' realities and practices of the sites, there is still retention of management control. Although the workforce is asked to follow the rules, the rules themselves remain. Where engagement and participation is sought, there is frequently still more a monologue than a dialogue of safety developed. Through this interweaving of enforcement and engagement, the traditional enforcement approach to safety management can still be found, albeit clad in the discourse of engagement that dominates modern safety management programmes.

Yet this interlinking of enforcement and engagement around safety is not necessarily beneficial and certainly does not result in a clearly defined understanding of what safety is, how it works, or indeed a coherent way of positioning safety within this context. Rather it serves to increase the diversity within the already disparate understandings of safety; developing constructions of safety as choice, safety as the rules, safety as following the rules, safety as punishment avoidance, all alongside the development of safety as engagement, ownership and personal responsibility.

However, an alternative reading of the enforcement/engagement of safety is able to consider the consequential segregation of these ideas around safety to be positive aspects of our construction site realities. Indeed, the hierarchy of management for the enforcement of safety it not just accepted by those working on sites, it is actually positioned as fundamentally *necessary* for the practices of safety and its management. Yet this also suggests the potential for conflict and dissonance in practice and process, and not least the potential for exploitation of engagement through the reality of enforcement. The desire for some form of clear management hierarchy from the operatives to keep them in check when they break the rules has not gone unacknowledged, supported by the reluctance of site-based management to totally cascade responsibility for safety down to the workforce. From the various hierarchical perspectives, and the subsequent positioning of the responsibility for safety, the hierarchy of safety at the very least supports the notions put forward by the Health and Safety Executive (2006) that clear and well-defined management roles are essential for the safe running of sites.

Summary

Examination of both the engagement and enforcement of safety provides deeper insight into management practices around safety, and the consequential shared understandings of how safety works in practice have been revealed. The variations within understandings that have been revealed up and down the site hierarchy are important, and it must be recognised that such differences in the way safety works exist throughout the site environment to inform the ways we try to manage and control safety through rules within these different contexts.

It is within this environment that safety propaganda seeks to manage, and indeed fundamentally change the process of management of safety on sites. Yet such developments are in sharp contrast to the traditional approaches of safety management through enforcement, which relied on the implementation of mechanistic regulations and vigorously enforced compliance (Langford *et al.* 2000; Haupt 2004). However, old management styles may have long shadows; the fundamental need for legislative compliance may necessitate some level of enforcement, if only to establish the standards and protocols to be met.

There is also the potential for dissonance and ineffectuality in the approaches made from the corporate level in their engagement with site-level realities. The lack of acknowledgement of the realities in which the safety management programmes are seeking to operate could develop barriers to their success, incompatible and incoherent within a reality of violation and the minor status afforded to such rule breaking. Indeed explorations of the enforcement of safety at the site level even suggest that hierarchical engagement and management is expected and welcomed by the workforce, indicating that ironically a more 'command' driven approach to safety as engagement would meet with success. It could be therefore suggested that safety violations must be accepted as everyday occurrence in order to ultimately eliminate them. Contemporary safety management programmes which seek to shift responsibility for safety to the personal could be incompatible with a workforce that feel control and punishment a necessary factor in their own safe working. Formal disciplinary processes and a zero tolerance approach are likely to still be required or at least acknowledged within the realm of safety engagement and the safety management programmes in order to create a paradigm shift in current thinking, and ultimately develop a mature site context which can then manage safety within a more violation-intolerant reality.

However, it is maturity of thinking that is critical here, and the most recent developments of safety engagement in the form of zero target safety management programmes have perhaps not sufficiently considered these wider issues of how safety works in practice – and so counting down to zero may not be as simply as we may think.

Acknowledgements

This chapter is based on work previously published in *Construction Management and Economics*:

Sherratt, F., Farrell, P. and Noble, R. (2013) Construction site safety: discourses of enforcement and engagement. *Construction Management and Economics*, **31**(6), 623–35.

References

Baram, M. and Schoebel, M. (2007) Editorial: safety culture and behavioural change at the workplace. *Safety Science*, **45**(6), 631–6.

Bernays, E. (2005) *Propaganda*. Ig Publishing, New York.

Biggs, H., Dingsdag, D., Sheahan, V.L., Cipolla, D. and Sokolich, L. (2005) *Utilising a Safety Culture Management Approach in the Australian Construction Industry* [Online]. Available: http://eprints.qut.edu.au/archive/00003797 [20 September 2015].

Dekker, S. (2007) *Just Culture: Balancing Safety and Accountability*. Ashgate, Farnham.

Dingsdag, D., Biggs, H. and Sheahan, V.L. (2008) Understanding and defining OH&S competency for construction site positions: worker perceptions. *Safety Science*, **46**(4), 613–33.

Frederick, J. and Lessin, N. (2000) Blame the worker – the rise of behavioural based safety programmes. *Multinational Monitor*, **21**(11), 10–17.

Gadd, S. and Collins, A.M. (2002) *Safety Culture: A Review of the Literature*. HSL/2002/25. Health and Safety Laboratory, Sheffield.

Hale, A.R., Guldenmund, F.W., van Loenhout, P.L.C.H. (2010) Evaluating safety management and culture interventions to improve safety: Effective intervention strategies. *Safety Science*, **48**(8), 1026–35.

Hartley, R. and Cheyne, A. (2009) Safety culture in the construction industry. In A.R.J. Dainty (ed.), *Proceedings of the 25th Annual ARCOM Conference*, pp. 1243–52. Association of Researchers in Construction Management, Nottingham.

Haupt, T.C. (2004) Attitudes of construction managers to the performance approach to construction worker safety. In S. Rowlinson (ed.), *Construction Safety Management Systems*, pp. 117–32. Spon Press, London.

Health and Safety Executive (2000) *Construction Health and Safety for the New Millennium, 313/2000*. HSE Books, Suffolk.

Health and Safety Executive (2005) *A Review of Safety Culture and Safety Climate Literature for the Development of the Safety Culture Inspection Toolkit, RR367*. HSE Books, Suffolk.

Health and Safety Executive (2006) *Successful Health and Safety Management, HSG65*. HSE Books, Suffolk.

Health and Safety Executive (2008) *Behaviour Change and Worker Engagement Practices within the Construction Sector, RR660*. The Stationery Office, Norwich.

Howarth, T. and Watson, P. (2009) *Construction Safety Management*. Wiley-Blackwell, West Sussex.

Hudson, P. (2007) Implementing a safety culture in a major multi-national. *Safety Science*, **45**(6), 697–722.

Hughes, P. and Ferrett, E. (2007) *Introduction to Health and Safety in Construction*. 2nd edn. Butterworth-Heinemann, Oxford.

Langford, D., Rowlinson, S. and Sawacha, E. (2000) Safety behaviour and safety management: its influence on the attitudes of workers in the UK construction industry. *Engineering, Construction and Architectural Management*, **7**(2), 133–40.

Lingard, H. and Rowlinson, S. (eds) (2005) *Occupational Health and Safety in Construction Project Management*. Spon Press, London.

Maloney, B. (2011) Conceptual model of safety culture for construction. In *Proceedings of the CIB W099 Conference Prevention – Means to the End of Construction Injuries, Illnesses and Fatalities*. CIB, Rotterdam.

Meldrum, A., Hare, B. and Cameron, I. (2009) Road testing a health and safety worker engagement tool-kit in the construction industry. *Engineering, Construction and Architectural Management*, **16**(6), 612–32.

Mohamed, S. (2002) Safety climate in construction site environments. *Journal of Construction Engineering and Management*, **128**(5), 375–84.

Palmer, L.R. (1926) History of the safety movement. *Annals of the American Academy of Political and Social Science*, **123**, 9–19.

Ridley, J. and Channing, J. (2008) *Safety at Work*. 7th edn. Butterworth-Heinemann, Oxford.

Wamuziri, S. (2011) Factors that contribute to positive and negative health and safety cultures in construction. In *Proceedings of the CIB W099 Conference Prevention - Means to the End of Construction Injuries, Illnesses and Fatalities*. CIB, Rotterdam.

Worthington, M. (2007) *The Behavioural Change Worker Engagement Forum* [Online]. Available: www.hse.gov.uk/aboutus/meetings/iacs/coniac/221107/m2-2007-3.pdf [20 September 2015].

Chapter Eight
Counting Down to Zero

The trouble with zero is there is no place to go ...

Long 2012

Yet despite this, zero has become the biggest number in construction site safety.

Many large construction contractors, and even some large projects, have made clear commitments to zero in terms of their safety programmes, and it's easy to see why. From moral perspectives zero is an absolute necessity. It is the only possible benchmark – no one working in construction sets out to have an accident, or to cause one for someone else. So no other number will do.

Yet whilst Hollnagel (2014) warned of the allure of the graphical presentation that propelled the accident pyramid to its continued prominence in safety thinking (as discussed in Chapter 4), zero should arguably come with several warnings of its own; the allure of the big round number, the allure of mathematics and measurement, and the allure of the snappy slogan – aptly illustrated in Figure 8.1.

Although seemingly simple in its goals and intentions, the use of zero within construction site safety may not be so straightforward when considered from a constructionist perspective, and placed with the inconsistent, incoherent and ever changing realities of construction site life. Indeed, how zero manifests on sites, and how it is associated with safety, has the potential to influence its

Unpacking Construction Site Safety, First Edition. Dr Fred Sherratt.
© 2016 John & Wiley Sons, Ltd. Published 2016 by John & Wiley Sons, Ltd.

Figure 8.1 Safety is No Accident

effectiveness in practice. Understandings of zero within the construction site context should therefore be unpacked to explore precisely how this biggest number in construction contributes to the understandings of safety and safety management now found on sites.

Target Zero: Theories and Thinking

As explored in Chapter 4, accidents and their statistics remain a key 'measure' of safety, and as a result this continued focus on the numbers has somewhat inevitably led to the need to 'Target Zero'. This approach also draws on understandings of cause and effect; if all causes can be identified then all accidents can be prevented and therefore zero accidents can be achieved (Hollnagel 2014: 63). Associating with ideas of safety as a non-event, zero therefore benchmarks safety in practice – no accidents making zero the true 'measure' of safety.

However, zero also has other associations. For example, a fundamental theoretical distinction needs to be made about the thinking behind zero. One of the most prominent manifestations of zero was the 'Zero Tolerance' approach by the New York Police Department in their tough stance on crime in the 1970s, which has subsequently been adopted, developed and implemented in many different contexts and industries. Indeed, within the construction industry – both in the UK and internationally – we now have zero carbon in the sustainability field, zero defects in our quality management, and of course zero accidents as the goal of our safety management.

Within the UK construction industry, zero accident cultures and programmes have become very popular, with zero proudly displayed on the hoardings of large construction sites and scrolling in banners across organisational websites. However, whilst a zero tolerance management approach can be readily put into practice by those tasked with safety management on sites, to set 'target zero' is actually very different. For example, for London 2012 the Olympic Delivery Authority adopted a 'zero tolerance' approach to unsafe practices and unhealthy working conditions on its project (Richardson 2006), which ultimately resulted in success. 12,000 people worked 80 million man hours in over five years to deliver the project, which was completed with an accident frequency rate (AFR) of just 0.15 and no fatalities on the project for the first time in Olympic history (Wright 2012). Although this project could well lay claim to zero fatalities, it did not achieve zero harm, or zero accidents; rather it achieved a highly impressive albeit realistic safety record, which was acknowledged as such by the Royal Society for the Prevention of Accidents with a special Diamond Jubilee Award. Yet under a Target of Zero the Olympic Delivery Authority would have had to carefully reflect upon their safety 'success', and perhaps even politely decline this acknowledgement and celebration of their efforts.

Although seemingly associated with key performance indicators (KPIs) and ongoing performance measurement found within many safety management systems, zero has subtly moved away from this approach. Instead, 'Zero Harm', 'Mission Zero' or 'Target Zero' often becomes the end itself, rather than any incremental steps on the road of continuous improvement. This is a fundamental shift in the underlying thinking of safety management, setting a milestone event to be achieved once and for all, rather than any accepted KPI measure of improvement, such as a reducing AFRs year on year. Whilst zero could have inevitably been assumed to be the ultimate target of such incremental improvements, the understanding that accidents will or may happen has now been set aside; no other number can be strived for.

Whilst the target of zero should of course be praised as the ideal, the way this has been repositioned within safety management, as the only ultimate and immediate goal, draws on ideas of enlightenment thinking – the belief that there will be a milestone event, an 'apocalypse' after which 'the flaws of human society will be forever abolished' (Gray 2007: 2). This necessarily also means that there will be a moment of identifiable and recognisable 'safety success'; the realisation of a target achieved, a goal attained, and after which we will work within a 'safety utopia' in which zero accidents, incidents or ill health will never occur again. When placed within the realities of industry organisation, projects and contemporary accident statistics, this seems something of a challenge to put it mildly.

Zero thinking around safety on sites tends to fall in line with what the historian Russell Jacoby (2005) termed 'blueprint utopias', where the future is mapped out in detail. This is something Jacoby considered to be a consequence of our 'image-obsessed' society (2005: xvii), where the propaganda around the utopian vision is more important that the thought behind the vision itself. This immediately starts to resonate with the logos and slogans that have come to dominate construction hoardings, hard hats and high-viz vests over recent years, and indeed the shift of safety itself from a site-based practice to a component of the social responsibility splashed all over corporate websites.

In contrast, a more emergent and organic process is also put forward by Jacoby, which he terms 'iconoclastic utopia', where dreams of a better society remain just that, rather than articulated and detailed representations. This approach allows thinking to escape what we considered to be 'normal' – the very shared understandings that make up our accepted realities, and instead asks 'what if …?'

This 'what if …?' approach is arguably far more useful to safety on sites, and would encourage questioning of many accepted and unchallenged developments around construction safety. For example whether the recent absorption of health and safety into the realm of corporate social responsibility (Rawlinson and Farrell 2010) is beneficial to its management in practice, or simply relegation in status? Whether the continued application of safety management systems to a production driven management system which neither supports nor fosters implementation in practice (Patel *et al.* 2012) is admirable perseverance or a foolhardy venture? These questions can and *should* be considered on a regular basis, and indeed this book attempts to go some way to explore them, yet this is very different to the carved-in-stone Target Zero. Instead, whilst there is inclination towards a better future, iconoclastic thinking refuses to map it out (Jacoby 2005: 85); its relevance is not to present practice, but a possible future where existing realities can be discarded for visions which encourage more fundamental change (Kumar 1991). An approach which would arguably lead to different thinking, and different ideas, rather than, for example, the repeated application of pre-determined, mechanistic safety management systems to our construction sites, places where accidents still occur.

There have also been more critical comments made of such utopian thinking in society. For example philosopher John Gray (2007: 28) feels a project is utopian if there are no circumstances under which it can be realised; if a project seeks to eliminate the fundamental contradictions of human needs it will not achieve success. Human needs in the construction site system arguably remain focused on time and money, where people are essentially paid to work faster and take risks to get the job done. Within the industry as it is presently structured, any health and safety improvements could certainly be considered, by Gray's definition, a utopian venture in the unachievable sense of the word.

It is the nuanced distinctions in approach and consideration of utopian visions that are most critical. Despite the need for construction health and safety management to consider its future 'utopia' from an iconoclastic position, to challenge current practice to catalyse change, it has instead developed a blueprint approach through zero. Drawing on the ever-prominent accident statistics, plans and targets have been employed, giving ideas 'a certain weight and plausibility' (Jacoby, 2005: 32) and providing the comfort of measurement, with focus on zero as the bull's-eye.

Targets are constructed as objects to be overcome, and which *can* be overcome, rather than the straightforward reflection of flaws in the ways of working that have historically developed on our sites. Hitting the target is itself hoped to bring about fundamental change, rather than change drawing us closer to the target. Zero accidents today providing reassurance that there will be zero accidents tomorrow. Its attractiveness and the readiness in which zero has been adopted by industry may reflect the fact that this is potentially a far more palatable process than the much more complex and challenging problems of instigating more fundamental industry change (around procurement, supply chains or methods and contracts of employment for example) as the necessary precursor to improved safety on sites.

However, the iconoclastic vision of Zero Target safety programmes is certainly not to be derided, indeed such intentions must be firmly supported – no other number would be morally acceptable. However, the development and positioning of these programmes within the context of construction site safety raises concern. Whilst a shift is certainly needed to move safety management forwards from its current plateau, setting Zero Target may not be the most effective catalyst. Whilst iconoclastic thinking should be championed, and the fundamental assumptions of *how* to manage safety revisited, the industry appears more concerned with blueprint incarnations of Zero Target, setting targets and steps within existing operational frameworks and practices. By their very construction, Zero Target programmes may inadvertently have a negative effect on the very environment they wish to change; the potential for disenchantment, disengagement, and their ultimate dismissal.

Brand Zero

Zero safety programmes on sites are often glossy, corporate and coated in PR sparkle. Indeed, they often create what in popular terminology would be a 'Brand Zero'; logos, slogans and other safety materials used to encourage the immediate recognition of zero within the organisation from the boardroom down to the site. Brand Zero can be found on websites, on posters and hoardings, on high-viz vests, the logo used to support every safety management tool used on the site creating a visual tag for safety management in practice, as seen on the corner of the signboard in Figure 8.2.

Brand Zero has generated a variety of bespoke safety programmes: 'Zero Harm', 'Mission Zero', and 'Target Zero' to name but a few. Even the titles of such programmes can be revealing. Zero Harm, for example, positions zero as the lack of negative action, although the use of harm is not restricted to health and safety, and indeed may be more immediately associated with environmental considerations in some contexts. Mission Zero and Target Zero instead make their zero a participant in an active process, with variations in the construction of that process as either a journey, a fixed point or commencement point.

Figure 8.2 Zero Harm

Alongside the big round number, a more intangible side to Brand Zero can also be found, framing zero as a way of 'thinking' and much more complexly, a 'culture'. Brand Zero can be used to either seek a cultural shift to assist in the attainment of the numerical zero, or to position the processes implemented under the banner of zero as critical in the development of a new culture. Yet this intangible and more flexible zero is still inevitably tied to the need for measurement – and zero remains the biggest of numbers. Such subtle variety in its corporate manifestations impact the influence of Brand Zero, and how it is able to integrate with wider safety management practice and understandings of safety on sites.

Some go even further – for example, the 'Beyond Zero' programme boldly announces on its webpage:

Zero incidents? We can do better than that!

But when this is unpacked, several different understandings of 'zero' in this context can be suggested. Firstly, '… better that that!' could simply be shifting the scope and challenge of zero beyond health and safety to include other areas of management practice – a bundling up of the non-productive (as explored and discussed in Chapter 6). And in this instance, 'Beyond Zero' does include environmental management, community improvements and quality considerations within the same text, gathering up these other elements within the remit of the zero 'approach'. This shared application of zero as 'thinking' rather than zero as a number or target is arguably an iconoclastic approach towards non-production, although health and safety is still reduced to the level of the bundle in terms of prioritisation and management.

Alternatively, '… better than that!' could suggest achieving zero incidents is an easy target. This reading was also identifiable elsewhere on the page:

> Aiming for zero accidents was a soft target and was not the final word in what could be achieved.

Here, Zero is arguably belittled beyond itself: positioned as 'soft', something so easily attainable that it should not be considered a target, just something to be looked beyond. When considered within the context of one of the highest risk industries in the UK (Health and Safety Executive 2014), this appears to be rather empty rhetoric.

A further reading could consider '… better than that!' as a desire to improve on the safety of the workforce, the text developing this construction further through associations with something 'actively positive' within health and safety management. This initiates considerations of positive liberty (Berlin 1958), and raises questions of where the corporate ends and private lives of the workforce begin. Yet such debates aside, zero has here been repositioned as the starting point, rather than the finish line of the safety management process. A big ask given the contemporary construction site context, and the lack of association with action or changes in practice to support such lofty ambitions.

But such hollow rhetoric is often an inherent part of Brand Zero – to make bold claims, to set the bar ridiculously high, to make commitments and pledges and promises that are signed up to by smiling executives in sharp suits and very, very clean PPE. It has become a victim of its own hype; the proclamation of zero alone is now suggested to assure organisational success and the very employment and use of Brand Zero has become equal in merit to any actual reduction in accidents, incidents or harm. Yet such proclamations of what can and will be achieved often fall short when considered within current industry contexts, and Brand Zero tries to position itself in practice.

Zero in Practice

The corporate iconography of Brand Zero does not simply cascade down onto the construction sites of the UK unchanged and unchallenged; the world does not work like that. In fact, significant variation can be found when Brand Zero is actually *talked* about on site; shared understandings shifting within the hierarchical positions in terms of their relationships with safety and practice.

At the site level, amongst operatives, Brand Zero loses its corporate glamour and becomes just another part of everyday site life:

> Well, you know, making sure everything that you do is done in a safe way.

In his explanation of a particular Brand Zero programme, this subcontractor's operative simply positions the programme as relevant to all work practice. Safety has become an inherent aspect of work – arguably the ideal – although this work and the responsibility for safety has been assigned to the more abstract 'you' rather than the 'I' of the speaker himself. But Brand Zero is not referenced in any particular way, the target or number is not seen to be of relevance, zero has simply become a tag for the more relevant – to this operative – realm of safe practice.

Even when Brand Zero programmes are specifically considered, zero is not a relevant number:

> It's eliminating hazards and reducing risks …

Although taking a more numerically grounded view, the elimination of hazards arguably reflecting zero in its approach, risks are simply 'reduced' rather than specifically reduced to *zero*.

For operatives, who do not have to toe a corporate line with perhaps the same vigour as their supervisors and managers, Brand Zero is dominated by the intangible: zero is not a relevant or necessary number and doesn't really have anything to do with their work – safety is safety, in one of its many forms and ways of understanding, but not a big round number. At this lowest level in the site hierarchy, Brand Zero is often just associated with practice and the realities of site work, rather than any tangible target or number, and is easily dismissed alongside the other slogans and logos of other the safety management programmes that have gone before.

Yet as the managerial hierarchy is climbed, amongst those with increased responsibility for safety, zero becomes much more prominent number:

> It's basically trying to reduce the accidents, the number of accidents, reportable or just incidents, down to zero.

This is much more reflective of the corporate voice; zero is the goal of the process of accident reduction. The speaker is careful to ensure that all accidents are included within their definition: both the number and type (reportable or incidents) of accidents are given importance. Zero is positioned as the ultimate target of the programme, a number to be achieved at completion.

As Brand Zero becomes more prominent it also becomes more complex, as managers and supervisors on site struggle to implement its stark numerical goal with the realities in which they work every day.

Indeed, the more abstract zero of the workforce can also be found amongst the managers and supervisors, who often try to position the numbers and targets within their own lived experiences and understandings of safety on sites. For example, to consider Brand Zero as:

> … more of a mind thing, getting people to change their mentality.

This may be in part reflective of the safety culture programme language of Brand Zero (the engagement approach as explored in Chapter 7), but it also moves away from the numbers, and instead moves back to safety and how safety works on sites, far more harmonistic with the workforces' own ways of understanding.

The hierarchy of safety has re-emerged and the position of the site management team as intermediaries between the site and the corporate levels will naturally influence their own understandings of zero. They must balance the relationship between the tangible and the intangible, the targets and visions and the realities of the site practice in which they must implement these programmes on a daily basis. There is a certain amount of incoherence and inconsistency in the form of a 'vision with targets', suggesting a desire to quantify a tangible zero, further developed and defined through the process of measurement itself. Such inconsistency does not negate either perspective; rather it is reflective of the shifting contexts and complexities of the realities in which the site management must work and the potential problems of implementing Brand Zero in practice.

Measuring Zero: Non-Accident Statistics

As the way people talk about Brand Zero suggests, the objectives and processes of Zero Target programmes are often unclear. A good example of this can be found in the *Vision Zero* programme, developed with the intention of improving road safety across Scandinavia and implemented in the late 1990s. Interestingly, the response to vision zero resulted in 'confusion as to whether (it) is to be seen as a concrete goal or more general ethical imperative (Swuste 2012: 1939). Differences in opinion could be found within the various countries seeking to adopt the programme. Norway considered Vision Zero to be an 'ethical foundation' and that 'the vision was not to be interpreted as a target' (Elvebakk and Steiro 2009: 958). In contrast, Sweden set themselves very ambitious targets including a reduction in fatalities in traffic accidents by 50% within a 10-year period.

Whilst the Norwegian response to zero is more akin to the way the construction workforce understand it, as a more intangible idea, the Swedish approach – the fixation on the numbers – is reminiscent of the UK construction industry, and the grand gestures made at the safety summit of 2001. At this summit, the Construction Industry Advisory Committee (CONIAC) pledged to beat the target set at the summit of a 10% reduction in fatal and major injury rates by 2010, and instead boldly committed to a reduction of two-thirds by 2010 (Health and Safety Executive 2009). Whilst the intention to improve safety on site was clear, although the construction industry achieved the 10% target, and continued in the same positive direction, it dramatically failed to make the giant leap to achieve the target it set itself. Despite the reassurances of measurement, *commitment* to a target does not mean automatic achievement, something often lost in the photogenic PR-friendly celebrations that often accompany their announcements.

Yet the UK is very fond of the targets and numbers that support blueprint utopian thinking, and these can be traced all the way back to Sir Isaac Newton. As noted in Chapter 3, whilst Newton's laws changed scientific thought, his clear mechanistic explanations cannot also be applied to the human and social worlds (Berlin

2002), but that has not prevented attempts to try. The most signifi-cant impact of this in the UK was the political misapplication of scientific game theory to society carried out in the 1990s, which created a highly target focused culture. But it did not work; people simply found new and ingenious ways to 'game the system', to meet targets through reclassification or reappraisal (Curtis 2007), and arguably in the UK our education, healthcare and other public services are still struggling to shake them off and fully recover.

For the UK construction industry accident statistics are some-thing of a sensitive area. Setting a target of zero could encourage sites to 'game the system', to reclassify incidents to meet targets, to simply under report, or to seek out alternative processes for measurement. This could of course be highly significant for the achievement of Zero Target in practice; construction work is an ongoing process and although each project has a definite time-scale, this does not necessarily translate to the wider organisation's statistical records, and so achievement at a higher level of opera-tions could prove problematic.

A further problems is the poor definition of parameters for suc-cess with regard to safety KPIs, which have been shown to create vague and irrelevant claims of achievement. For example, the claim that a safety KPI was 'broadly achieved' (a 'measure' identified by Rawlinson and Farrell (2010) in their review of safety targets in practice) is certainly no longer acceptable when the target is a very clear zero. As found in other applications of measurement to the potentially immeasurable, additional targets and more complex management controls are likely to be needed to support the initial target (Curtis 2007). Complex, unwieldy and time-consuming measurement systems and reporting procedures can develop, nec-essary in the attempts to measure the deceptively simplistic target of zero.

A focus on numbers can also prove problematic in terms of the practical achievement of a safety 'apocalypse', the moment when the Zero Target has been achieved. In their aim to eliminate all accidents, Zero Target programmes are stating that 'accidents won't happen'. However, as shown by the challenges made to sim-plistic 'cause and effect' thinking, it is misleading to suggest that all accidents can be prevented (Hollnagel 2014) or that the probability of an accident can ever be reduced to zero (Whittingham 2004: 254). Brand Zero is therefore challenged by more contemporary theories of the nature of accidents, relying instead on more out-moded ideas of causality, as well as their facing challenges from the fact that accidents remain all too commonplace occurrence on sites, and the ingrained belief that they will inevitably happen.

Indeed, responses to the Scandinavian Vision Zero suggested the 'zero of Vision Zero' was 'unrealistic' (Elvebakk and Steiro 2009: 963). Such thinking may lead to instant dismissal of the programme as a consequence of its perceived unattainability. In setting a blue-print utopian target of zero the entire vision becomes vulnerable; as a simple consequence of just one accident the target becomes unachievable, it is lost, potentially disenchanting those it is seeking to inspire. Arguably, the approach falls foul of its own construc-tions. Zero Target programmes could be suggested to begin from an untenable position as defined by their own terminology.

Achieving Zero

In UK construction, the achievement of zero has been positioned variously against various methods of quantification, including time-scales, targets and ratings.

Measurement by time constructs zero as a tangible target with a future year assigned to the achievement of zero by the corporate voice, for example the first incarnation of Balfour Beatty's Zero Harm promised this:

> … by 2012

But setting a date for safety success can become nonsensical with-out further explication. Is the target is to be implemented from the target year onwards? Or from a fixed point before in order to 'achieve zero' at this date? This makes time highly problematical in terms of parameters, measurable criteria and ultimately interpretation. Although this fuzzy articulation could also be deliberate, construct-ing associations with ongoing improvement and progress necessary to meet this target in the future, creating a reality in which action is needed now, to encourage and instil support for the programme within the workforce, without further articulating its measurement.

But using time as the measure does of course become a big problem should the year be reached and the target not be achieved. Despite making significant improvements in overall safety perfor-mance, Balfour Beatty did not achieve 'Zero Harm by 2012', and indeed have since revised its programme, omitting any timescale from its strapline.

Whilst measurement by time has the potential to create clashes with reality upon its inevitable arrival, it can also construct commitment through the future positioning of the target to a point in time where reality may have changed to one where zero is attainable, if not the norm. This resonates with the more intangible understandings of zero, articulating a target in the future with time to work towards it, whilst continuing with everyday practice.

> At the time you think it's a massive ask, but we've got years to achieve it in ... especially the way our accident figure have gone since we started.
>
> It's just a further step change from there ... more of a mind thing.

This site manager is using the timescale itself as a facilitator in the achievement of zero, reflecting on what can and already has been achieved, supporting a notion of success over the duration of the programme rather than any immediate change. Zero is constructed as a challenge, but not one that is unachievable as associated with current practices and successes. Despite this consideration of zero as 'accident figures' the speaker ultimately considers zero to be a 'mind thing', shifting towards intangible understandings of zero. This variation in the constructions of zero is again reflective of the complex relationships between the tangible and intangible zeros, and the different and shifting ways people understand and position zero within their everyday work practice. Although the speaker is able to measure zero through both time and accident quantification, there is also a reluctance to focus solely only on measurable targets, and the reference to zero as intangible makes wider associations with the more immeasurable people and their practice.

A more simplistic approach is often found in the corporate zero, where measurement by numbers alone is much more common. Targets and numbers are embedded within the discourses of performance improvement and management, and are associated both with the ongoing process of zero as well as zero as a tangible target. Ongoing zero is often constructed through associated criteria, such as incremental reductions in accident frequency rates (AFRs), indicators, scores and KPI targets positioned as a measurement of the impact made against zero. This approach makes zero something larger than the targets themselves, a solid and tangible entity, and the programme an attritional process in achieving the wider aim.

Numerical targets are also used to demonstrate the progress of the organisation towards zero. For example the statement that:

84% of (company name) contracts achieved zero reportable accidents.

puts this organisation's attainment of zero within some readily understandable constraints. Yet reportable accidents are those notifiable under the Reporting of Injuries, Diseases and Dangerous Occurrences Regulations 2013, which does not include *all* accidents, incidents or indeed harm. So whilst this claim does not therefore truly reflect the zero target put forward by the organisation elsewhere, zero here has instead been packaged up to the project level, creating a parameter for zero that makes it more achievable (well, 84% more achievable) in much more incremental and indeed generously recorded measures.

Alternatively, zero can itself be constructed as the target, with companies variously committing to:

achieve zero harm by 2012

by 2020 we will eliminate all accidents, of any severity

reduce our all AFR by 70% by 2020

All of these pledges create a shared sense of corporate ownership of the target, with zero positioned within the parameters of both time and quantification. Yet as previously noted, such constructions can fall foul of the use of time in their measurement, and they can also suffer from a single challenge to their numerical quantification. Positioning zero as a tangible, numerical target creates the potential for just one accident to bring failure and cancel its attainment, as a site supervisor explained:

Obviously you've got the odd project that something happens, and it ruins the figures.

By associating zero with numbers and measurement, the supervisor quantifies his evaluation and positioning of difference within site practice. But by considering measurement as success, there is now the inevitably of 'ruin' should an incident occur. What this 'something' could be is not really clarified, and makes no evaluation of the incident in terms of severity or potential consequences to a worker in reality, rather the focus remains on numbers and measurement of safety.

Something stupid happens, and you've lost the target.

The supervisor now develops 'something' but this again makes no recourse to the incident in terms of practice or people, rather it is constructed with direct reference to the impact on the target of zero, and belittled, considered 'stupid'. Whilst the safety incident is positioned as a negative event, it is only considered as negative through its direct association with Brand Zero.

It can be suggested that a consequence of this focus on Brand Zero has produced a depersonalisation of health and safety management. Within a project or organisational context, one individual alone is not able to prevent all accidents; it requires shared practice if only because of the logistical and physical demands of the site. Yet under Brand Zero, this shared focus can be easily affected by others, and so responsibility for success resides within the shared workplace, which in turn can lead to individual disassociation with health and safety management at a personal level. Just one incident can therefore 'ruin the figures' and potentially disenchant those who should still be focused on safety management in practice.

Associating zero with measurement inevitably holds repercussions for the attainment of success. Parameters of time or quantification construct zero as something tangible which can either meet with success or failure. In contrast, zero in its intangible form arguably cannot fail or succeed, yet the everyday associations of measurement with any number make this evaluation to some extent unavoidable.

More significantly, as soon as the measurement of zero meets the 'muck and bullets' of construction site reality, it becomes highly problematic for those who work there:

> You could never get rid of human error … people are working long hours and they're tired … you'll never get rid of injuries.

This gateman has associated zero with practice, and in doing so has positioned zero firmly within his own site reality in which people are the cause of unsafety, and the reason zero cannot be achieved. But this is not blame as prescribed in the simplistic causal thinking of human error and accidents, rather more systemic causes are naturally identified, long hours and fatigue, creating a close relationship between zero–people–work. For this worker, zero in its tangible form is simply unachievable; the current reality will dominate and injuries can never be eliminated.

That construction sites involve people, and people are inconsistent and irrational, as social constructionist theory suggests and supports, is often revealed when zero is considered:

> People are the unknown, some days they will and some days they won't.

For this operative zero has no chance when placed within the realities of practice, and specifically people. People are the 'unknown', the reason for the unachievability of zero.

This shared understanding of construction sites frequently emerges through talk of zero; that it is people and practice that make a tangible target of zero unachievable. Safety management, legislation, PPE or any other manifestations of safety in practice are not necessarily lacking, rather there is something fundamental about the 'nature' of sites themselves that is problematic, clearly linking back to the discussion of safety versus productivity explored in Chapter 6.

However, despite the general acceptance that zero can never really be achieved, this does not negate its necessity as the target, and iconoclastic visions of zero can also be found on sites.

> We won't do it but if you aim for it all the time I suppose you'll get closer.

Although resignation weighs heavy here, and although zero is unachievable within *current* practice, the target of zero was not itself derided by this operative. Rather, zero is the acceptable target to be aimed for, rather than be achieved in itself, it is undefined steps on the journey towards it that are more realistic and attainable within the existing reality of the site environment. This is a truly iconoclastic utopian vision; zero becomes a necessary goal for the workforce on site and is reflected throughout by those who have to put zero into practice. On sites zero has become the accepted ultimate target, with no recourse to the events that might realise this in practice, and indeed no belief that it will ever actually be realised, but it remains an optimistic daydream just the same.

But the rejection of a tangible achievement of zero in practice found on sites is in sharp contrast to the corporate Brand Zero, which very clearly sets target zero amongst a variety of temporal and numerical parameters. It is the application of measurement to zero that makes it something tangibly achievable, either the ultimate target or the incremental measured steps towards it. But this web of measurement forms the framework of Jacoby's (2005) blueprint utopias, secure in a world of quantification, yet reluctant to think beyond current practice to seek out significant change.

So whilst corporate Brand Zero operates within a controllable and manageable site context, where specified practices could be easily implemented to support the achievement of zero, operatives struggle; the clinical, clean zero and the complex, confused site environment is not easily reconciled within the context of everyday practice. Whilst zero is deceptively simple, people are simply complicated, and the practice of work on site is highly influential in the construction of a reality of unsafety, in which zero remains a number that can never be attained.

However, this mythical status does not completely negate the value of zero in practice; no other number will do. Despite concerns with measurement, despite the potential for failure, despite the hostile context and despite the fact it is 'unachievable', zero is likely to remain an essential component of health and safety management on sites – and is perhaps something that should form one of the building blocks of our industry castle in the air.

Summary

The emergence of Brand Zero in construction safety programmes arguably reflects a wider societal desire, in the UK at least, to quantify and measure human life. Following the shift in political thinking of the 1990s and the application of scientific game theory to practice, target-driven systems have become the norm. When health and safety on UK construction sites is considered, ethical and moral concerns become paramount and in an industry where corporate social responsibility has developed into a significant marketing and pre-qualification tool for commercial success, no number other than zero could ever be acceptable.

Consequently, the corporate voice of Brand Zero speaks of an achievable tangible goal, positioned as a firm future reality, which can be counted and measured through a plethora of targets, and even gone beyond to bring positive health and safety to the workforce. Yet this blueprint utopia is challenged and even derided by the construction site workforce, who take an iconoclastic position. As a result of this, zero is considered an unachievable target, something incompatible with the current challenges of practice they face on a daily basis, and so they are instead simply content to look with shared ownership towards a potentially brighter future.

It is the desire for measurement that brings zero into its ugly reality. Such blueprint utopian thinking does not seek to challenge and change current practice; rather it aims to operate within the existing environment, seeking engagement of the workforce without addressing problems of practice. Furthermore, associations with measurement have also encouraged a focus (or rather a distraction) on the numbers and continuous improvement, rather than the practices and the people behind them.

There is the potential for Brand Zero thinking to bring change to health and safety management on UK construction sites, but simply setting a target of zero is not itself worthy of celebration or indeed merit. Arguably a shift in focus is needed from the miniature, the numbers and the 0.01% improvements, to look to the bigger picture. To ask why practice is so prominently positioned as a challenge to zero, to ask why people who spend each day on UK construction sites are happy to position zero as a vision for the future, but remain derisive of its achievement within their current working contexts? The present target of Brand Zero is arguably hindering its own achievement, providing distraction and comfort in the application of numbers and mathematics to something that is actually about the complex, awkward and immeasurable world of people and practice.

Acknowledgements

This chapter is based on work previously published in *Construction Management and Economics*:

Sherratt, F. (2014) Exploring 'Zero Target' safety programmes in the UK construction industry. *Construction Management and Economics*, **32**(7–8), 737–48.

References

Berlin, I. (1958) *Two Concepts of Liberty – An Inaugural Lecture delivered before the University of Oxford on 31 October 1958.* Clarendon Press, Oxford.

Berlin, I. (2002) *Freedom and its Betrayal, Six Enemies of Human Liberty* (ed. H. Hardy). Pimlico, London.

Curtis, A. (2007) *The Trap: What Happened to Our Dream of Freedom – The Lonely Robot.* 18 March 2007. BBC2, UK.

Elvebakk, B. and Steiro, T. (2009) First principles, second hand: perceptions and interpretations of vision zero in Norway. *Safety Science,* **47**(7), 958–66.

Gray, J. (2007) *Black Mass: Apocalyptic Religion and the Death of Utopia.* Penguin, London.

Health and Safety Executive (2009) *Underlying Causes of Construction Fatal Accidents – A Comprehensive Review of Recent Work to Consolidate and Summarise Existing Knowledge.* The Stationery Office, Norwich.

Health and Safety Executive (2014) *Health and Safety in Construction in Great Britain, 2014* [Online]. Available: http://www.hse.gov.uk/statistics/industry/construction/construction.pdf [1 October 2015].

Hollnagel, E. (2014) *Safety I and Safety II – The Past and Future of Safety Management.* Ashgate, Farnham.

Jacoby, R. (2005) *Picture Imperfect: Utopian Thought for an Anti-Utopian Age.* Columbia University Press, New York.

Kumar, K. (1991) *Utopianism.* University of Minnesota Press, Minneapolis.

Long, R. (2012) *For the Love of Zero: Human Fallibility and Risk.* Scotoma Press, Australia.

Patel, M., Sherratt, F. and Farrell, P. (2012) Exploring human error through the safety talk of utilities distribution operatives. In S.D. Smith) (ed.), *Proceedings of the 28th Annual ARCOM Conference,* pp. 403–12. Association of Researchers in Construction Management, Edinburgh.

Rawlinson, F. and Farrell, P. (2010) UK construction industry site health and safety management: an examination of promotional web material as an indicator of current direction. *Construction Innovation,* **10**(3), 435–46.

Richardson, S. (2006) How will this man make the Games safe for workers? *Building Magazine,* Delivering 2012 Supplement, November.

Swuste, P. (2012) Editorial – WOS2010, on the road to vision zero? *Safety Science,* **50**(10), 1939–40.

Whittingham, R.B. (2004) *The Blame Machine – Why Human Error Causes Accidents.* Butterworth-Heinemann: Oxford.

Wright, E. (2012) Olympic health and safety: record breakers. *Building Magazine,* 1 June.

Chapter Nine
Constructing Safety on Sites

... this is where the concept of safety culture comes into the picture.

Choudhry et al. 2007: 1003

So what has this unpacking of construction site safety achieved? Can we now answer the simple questions posed in the introduction to this book?

Well, to some extent we can, but the answers are not easy ones. Throughout this book there has been a theme of change and inconsistency, both within the construction site environment itself (where bringing changes to spaces and places is something fundamental to what we do) but also within the people that work there – in taking a social constructionist view of reality we have accepted that people are not consistent, coherent, or even particularly rational. But this acceptance of the inconsistent and incoherent has opened up many different ideas about safety – we have not tried to seek definitions or create lists of how to 'do' safety on sites but we have used these ideas to unpack safety within this highly complex context, through a lens of changeability, which has revealed some of the key understandings that are shared around safety on site. We have gained some useful insights, and one of the most important is probably that there is no easy answer to the first question posed:

What is safety on site?

Unpacking Construction Site Safety, First Edition. Dr Fred Sherratt.
© 2016 John & Wiley Sons, Ltd. Published 2016 by John & Wiley Sons, Ltd.

Safety is actually highly complex, and whilst that might sound like academic nonsense, it is arguable that this simple realisation alone is itself pretty helpful.

If we sit back and just think about how complicated safety is – how it is different for different people, how it shifts and changes in its relationships with work, whether it is embedded with in practice or left to stand alone, or even pushed off to one side with all the other non-productive aspects of construction management, whether it has a fixed state or is far more intangible and shifting with time and space, whether it is linked more closely to management and control, engagement and enforcement, ownership and responsibility – we can start to get a better idea of how complicated its management *should* perhaps be.

Safety isn't something that can necessarily be managed by ticklists or generic inspection forms created by off-the-shelf safety management systems, indeed the ways we seek to measure and manage safety have been revealed within this book to be at times much less than helpful themselves.

Safety is awkward, inconsistent and even paradoxical at times, but knowing this means we are much better placed to understand it, and to do something about it.

This last chapter will seek to bring together what this book has shown construction site safety to be, and then seek to explore this through the familiar label of a 'site safety culture' – considering whether this is a helpful approach or simply yet another complication in the management of safety on sites. Finally, this chapter also aims to ultimately develop some practical outputs from the new understandings and illuminations gained through this book, to try to support the development of effective interventions to improve construction site safety in practice.

What is Safety on Site?

Safety on construction sites operates within two distinct contexts. Firstly, the wider processes of the construction industry arguably create an environment in which safety struggles to survive. The constant pressures for production, as time and money are prioritised through the practices of lowest cost tendering and ever-squeezed margins along protracted and convoluted supply chains, are highly influential and can be readily identified as fundamental 'truths' of construction site life. On sites, the presence and inevitability of change becomes another truth; the need to build and develop our working spaces around us is an essential part of the job, yet change is not something that is easily managed in practice, and is something that has considerable influence on safety. That construction sites also frequently contain challenging working conditions, such as dust, noise and exposure to the elements, also contributes to an immediate workplace context that is not necessarily embracing of or even that receptive to safety in practice.

When we try to position our sites within the wider social understandings of safety, things get even worse. In the UK at least, safety is derided and ridiculed throughout society – we live in constant fear of a 'compensation culture', restricted by a rapid growth in ridiculous rules and regulations labelled in the 'interests of safety' (Brown and Hanlon 2014), and through the innocent, although at times much more deliberate, misunderstandings of safety frequently shared throughout our media. Yet this means safety is also derided in our workplaces, and this includes our construction sites. Macho posturing by some construction workers has not helped matters, as they share their daring exploits and pranks through social media and websites like *On the Tools*, laughing at safety (or rather unsafety) on sites in a way which beautifully sums up not only how it is positioned in our wider society, but also within construction site practice.

It is within a combination of these two contexts that we attempt to manage safety on sites, and which aspects hold most influence will inevitably change on a regular basis. We acknowledge the news of a fatal accident on a site in another city by sharing stories of our own experiences and expressing our sorrow for the families of those involved, but we also attend our morning meetings where deadlines are carved in stone and the expense for extra equipment that might help us work more safely simply won't be forthcoming. We attend the tool-box talk and try to pay attention to the fact that the access routes round site will be changing this week because of the floor screed works starting, but we also have a great laugh in the canteen at watching the latest video of two lads trying to get a roof truss up on the scaffolding and the whole bloody lot coming down on top of them.

Change is something that has been acknowledged throughout this book – both change in the contexts of construction site life, but also change in the people that work there. Our constructionist approach has shown them to also be inconsistent and incoherent, which causes further problems in answering the question *what is safety on site?*

In the first place, this challenges our ability to *agree on our definition* of safety. And to answer this question as it was posed in the introductory chapter to this book – in a word, no, we don't agree, and it is actually debatable if we ever could. Safety is many different things to different people – as a social construction, safety itself is in constant flux, although we have been able to explore and identify some of the key ideas that surround it throughout this book. For example, the most common way we talk and understand safety is through PPE, the most prevalent and (perhaps inevitably) highly visible idea of safety on sites, despite it only being an artefact of safety management. But PPE is what we all too often think safety *is*, and of course it's not. But it is easy to see, easy to check and easy to record compliance on the site inspection check-sheets – and it could be argued that if PPE is OK then we can assume everything else is too. But then someone standing on a leading edge or the back of a wagon becomes safe as long as they have their hard hat on – never mind the fact that they might fall.

We also struggle to separate safety and unsafety. A reliance on accident statistics to tell us about safety is actually giving us a measure of unsafety in practice – safety is itself a non-event and therefore as Townsend (2013) has so vigorously argued it 'can't be measured'. Yet we still try, and are even trying down to the ultimate goal of zero. In contrast we *do* look to manage safety through our safety management systems, but this places safety outside of the business of production and segregates it from other management processes. Perhaps we would be better to try to manage risk and danger in practice, and then safety could take care of itself. We are also inconsistent in our approach in applying the labels of safety or unsafety on our sites – signs often proclaim 'Danger!' precisely on the measures put in place for safety, yet they also shout precisely the same thing in places where no measures of safety have been applied. Constructing safety as danger is paradoxical, but as we have seen, this could easily have developed through the ways we understand and try to define safety – the need to construct safety as the absence of danger has led to ideas that sites are inherently unsafe places and sources of constant danger, even when such danger is being managed and controlled.

So what is our definition of safety? It is one that involves unsafety and accidents, it is one that sees safety as danger, but it is also one that is happy to define safety as safe through the ways we seek to manage it in practice. And to a large extent *how we talk about safety* has influenced this development. Grounded in the legal lexicon and the many regulations that direct safety management in practice, ideas of safe/unsafe have come to dominate the way we can talk about safety. This has clearly had influence within our safety management systems and the labelling of safety – to proclaim safe is often highly complicated and too challenging for many (particularly with the compensation culture of the wider social context lurking in the background) and so it is often easier to simply proclaim and label unsafe. Yet whilst just two safe/unsafe boxes look lovely on the site inspection check-sheet, they don't really work in practice.

Instead, out on site, this has been overcome by the dismissal of such polarised ideas of safety – and instead safety is something fluid, a moving target, and for things to be 'a bit unsafe' is a perfectly acceptable way of understanding safety. Safety is something that can change, it can develop and it can shift quite easily – not too surprising given the ever changing context of the site itself and the shared understandings of those who work within it on a daily basis.

But nevertheless, we do still try to embed the polarised safe/unsafe into practice, often through formal safety management systems, but this has often meant that safety or safe is totally removed from work. With no way of easily fitting alongside production, safe becomes disassociated, an entity in its own right, something that has its own influence – and is therefore differently *associated with practice*. Indeed, safety as an entity can influence work practices, often in the negative – stopping the job, causing delays and generally getting in the way. In making safety an entity, it can also be all too readily bundled up and put to one side with all the other aspects of construction that do not directly contribute to production – health, the environment and quality often become unlikely bedfellows, yet one poor manager is often tasked with meeting all these requirements. Safety becomes equivalent to the need to ensure all the timber on the project is from sustainable sources (no matter how long the supply chain) or the need to plant an allotment for the local school children using recycled topsoil – safety at risk of becoming another opportunity to demonstrate corporate social responsibility, despite the fact that for safety, compliance is actually required by law. But although safety is often positioned as an entity, it can also be found embedded in practice, people do understand safety to be an inherent part of their work, and it is then that work practices negatively influence safety – making it hard to work safely when the pressures of production come into play.

So, drawing on these varied, convoluted and often contradictory understandings of safety, we need to ask a very fundamental question – *Does it 'work'?*

Well, not really. These shared understandings of safety have influenced the way safety works in practice – including the critical considerations of responsibility and ownership. A struggle can also be identified here. Operatives seek out safety enforcement, with rules, violation and punishment required and even vital for the safe operation of sites, as they try to operate within the complex contexts and ever changing prioritisations of safety in practice. Yet corporate management often ignores violations and punishment, and instead seeks a no-blame culture where responsibility for safety is cascaded back down to the operatives themselves through personalisation and ideas of shared ownership for safety. Clearly these two worlds do not come together well. And the site manager's daily struggle to bring them together is easily revealed, through handwritten threats of punishment pinned to the canteen walls, suggesting that trite slogans and superficial personalised communications around safety are not proving particularly effective in practice. And the most recent step towards Brand Zero has added further complexity to the mix – and again revealed these worlds as incompatible – whilst corporate management aim for zero in practice, the workforce instead are happy to build their castles in the sky, content to aim for zero, but with the firm resignation that it is never going to be actually achieved for safety on site.

So if current ideas of safety on site are not really working, what can be done? Well, as the introduction to this chapter suggested – the ideas and understandings of safety unpacked above are indeed helpful. They are able to illustrate what safety *is* on site – and that it is all a bit of a mess is itself well worth knowing. However, it would perhaps be more helpful to be able to understand these ideas in a more coherent way – and whenever the complexities of both people and the social worlds are explored, very often the concept of culture comes to the fore. This is what Choudhry *et al.* (2007: 1003) were alluding to in the quote found at the start of this chapter. Where complexity and dissonance start to emerge around safety, the check-sheets and tick boxes of proceduralised safety management systems become less relevant and instead alternative conceptualisations and underlying theories of culture often emerge, focusing on human and social aspects through ideas of safety culture.

Site Safety Culture

As noted in Chapter 7, it is generally acknowledged that culture is one of the most problematic concepts within the English language. However, it must also be recognised as one of the most enduring – perhaps because of its more intangible qualities, which to some extent let us off the hook in terms of any management or control. Within the construction industry it is very unlikely that the idea of a 'site safety culture' is going to go away anytime soon, and so it is worthwhile exploring how this might work in practice, and indeed how any intangibility might help us better conceptualise the ideas around safety revealed in this book.

Safety culture within the construction industry is often approached in the same way as safety itself – through safety culture management programmes and systems that again necessitate the presupposition that safety culture *can* be managed and changed for the better. This has in turn influenced the way safety culture is conceptualised and consequently employed in contemporary management practice. Details of how safety culture fits in with current safety management practices were explored in Chapter 7, through the ways it seeks worker engagement through leadership, communication, a no-blame culture and programme identity, all of which are seen as contributory factors in the development of a positive safety culture on sites (Wamuziri 2011).

This type of approach is firmly rooted in a normative conceptualisation of culture, where culture simply becomes the manifestation of the formal procedures, policies and structures that shape a social environment which can be assessed and developed against a prescribed 'best practice' (Edwards *et al.* 2013). Such organisational policies are relatively simple to review; pledges for management commitment and the presence of a no-blame culture within the organisation can be stated and signed, and published as part of a revised safety culture focused policy. Procedures and practices are also relatively easy to adapt; for example inductions for all site operatives onto the safety management programme, and the establishment of the relevant committees and representatives from among them. Such developments can then be monitored on an ongoing basis using safety climate surveys, to measure and quantify the resultant safety culture in practice (Guldenmund 2007; Fang and Wu 2013). A further benefit of this 'to-do list' approach to safety culture is the fact that it creates things that can be measured. As we have seen, this is something highly attractive to safety management in practice, despite the fact that research has found safety climate and safety performance to be 'weakly' related at best (Clarke 2006).

Nevertheless normative conceptualisations of safety culture have been readily adopted by construction organisations, arguably because of the good 'fit' with current management practices, enabling organisational 'cultural change' to be brought about with minimal effort. The 'measurement' of safety can be easily incorporated within the site manager's weekly inspection forms, and subsequently assessed against contemporary safety key performance indicators to measure safety culture within the organisation or on the site.

However, such approaches to safety culture have been challenged. For example, Dov (2008) identified significant discrepancies between such formally declared safety policy and actual safety practice, whilst Choudhry et al. (2007) also found that construction organisations' safety management systems as they exist on paper do not necessarily reflect the way work is carried out in practice. Although Antonsen (2008) suggests that this dissonance may be the result of the development of policy away from the workplace, which can create unworkable practices and subsequently necessitates the development of informal procedures which do not conform to the same safety standards, it could also simply be a consequence of the way safety works. As this book has revealed, if policy is built upon the formal legalese of safety, where the only options are safe/unsafe, then such policy will *always* struggle to operate within the fluid and flexible understandings of safety found in practice, as well as pose problems for management as they try to align the simple directions of safety policy with the inconsistent ideas of safety that form part of their own lived realities. The polarisation found at the root of policy might also further serve to disassociate safety from practice, through the creation of a separate and segregated safety culture, rather than any incorporation and inclusion as part of a 'culture of working safely'. Again, the management focus on developing a strong 'safety culture' has supported the construction of safety as an independent entity, and in fact the term 'safety culture' has long been considered inappropriate by some (for example Hale 2000 and Hopkins 2002) for precisely those reasons.

To follow contemporary construction industry practice, and retain focus only on the normative 'measures' of culture is clearly not helpful, given the understandings of safety that have been revealed in this book. We need to look beyond this highly limited approach to culture. Debates around what culture is are ongoing in a wide variety of academic fields, with differing approaches and therefore different definitions and conceptualisations. For example, culture can be seen as something that is shared and created by people within and through their social interactions, but people are also shaped by the culture in which they are born, live and work, because culture provides individuals with the ways in and through which they understand their social environments (Antonsen 2008). Culture cannot therefore be a homogenous entity or tangible 'thing' – it is multi-layered and multi-faceted (Dov 2008), consisting of different subcultures operating within their own social contexts as well as interacting with each other in wider social settings. It is also constantly shifting and developing with changes in context, as individuals or organisations join or leave the social settings in which shared meaning is created.

This is of course also very relevant to the ideas of safety that could be seen to contribute to a different kind of safety culture in which understandings of safety emerge not only through lived experiences, but also from the wider positioning of safety in society. Indeed, 'safety culture' has long been accepted as a social construct (cf. Denison 1996: 636; Antonsen 2008: 24; Glendon 2008: 250), as safety itself is constructed and reconstructed on a daily experiential basis (Guldenmund 2007: 741). This version of culture – one that is multi-faceted, multi-layered, shifting and changeable entity – fits very well with what we have revealed construction site safety to be throughout this book, and how it actually works in practice.

Safety culture, when considered from the constructionist perspective reflects Dov's (2008) multi-cultured and multi-layered construct. Through the shared understandings of construction site safety differing layers of 'corporate culture', 'management culture' and 'operational culture' have been revealed, each operating within its own parameters of rules and punishment, through the engagement and enforcement of safety. Safety culture is fluid, shifting along its own continuum from safe to unsafe and through many different constructions and manifestations in between, as well as varying in levels of association with both the practice of 'doing safety' and the practice of 'doing work'. Normatively, safety culture operates only at the ends of this continuum; safe/unsafe providing the formal policy and procedures by which safety can be readily measured, yet as a consequence of how safety actually 'works' on sites, only one of these extremes, unsafe, dominates in terms of recognisable management practices. The roots of these constructions have also been revealed; the lexicon of safety and its development though legislation and management practice have constrained safety into a polarised form that is incoherent and uncomfortable within the ever changing construction site environment.

So is this more flexible and experiential version of safety culture useful and helpful for those trying to manage safety on sites?

It is perhaps reassuring to boldly state and accept that safety culture can't be measured, and more fundamentally neither can safety itself. This relieves us of the need to develop ever more complex 'measures' of safety, more convoluted forms and inspection sheets, more detailed policies and procedures, and the bureaucracy that often sits alongside such paper-based approaches that reflect normative concepts of safety culture that don't really work in practice. Instead, this version of safety culture enables us to better prioritise the individual and social aspects that are inherently involved. How people *understand* safety is important; how it is developed, associated and shared by those interacting in the work environment, what they consider significant in their actions and interactions. But we must remain mindful of the fact that this is not something that can be measured either. It is people that contribute and ultimately create the changeable and complicated version of safety culture in practice found on our construction sites.

And in part this summarises what this book set out to do. To reveal 'safety culture' for what it actually is – something complicated, changeable and inconsistent. To remind us that it is people that create a culture, not check-sheets, policies and sophisticated safety propaganda. It is people that make our safety culture on sites, and unfortunately people are complicated – as changeable and inconsistent as the construction site itself. Safety culture is all too often used as a catch-all phrase within safety propaganda or management systems with no acknowledgement of the difficulties and complexities that are actually behind it. Instead normative ideas dominate, and we cling to things that can be measured in reassuring ways. However, this approach is not really helpful for managing safety on sites, and has actually contributed to the perpetuation of an understanding of safety that simply does not work in practice.

So how *can* we better manage safety? Well, perhaps firstly we need to shift away from managing safety and look instead to how we manage people and their cultures; we need to rewrite 'safety' with greater reference to actual practice and lived experiences. This means we need to challenge and change the way people understand safety – but, as we have seen, management influences or actions do not necessarily cascade throughout our sites unchallenged or unchanged. However, this book has revealed and illuminated some areas that could instigate positive change, and could help to make safety work better within the construction site context.

Putting it into Practice

It is hoped that this book will help and inspire those whose work directly involves construction site safety to rethink how we manage safety in practice – they are most readily able to use these ideas and different understandings of safety to help develop or inform interventions that have a much better fit with the realities of safety as found in everyday construction site life. They can review existing safety management systems and safety practices, and unpack how safety is positioned and talked about in their organisations, and assess fit and relevance from these different perspectives. This book has hopefully been able to illuminate why some safety interventions work and some don't, and consequently puts forward the basic premise that *any* safety intervention must accept the ways safety works on sites before they can seek to change it. We need dynamic and responsive safety interventions which acknowledge that safety isn't an entity – something separated from work and practice and sat on its own in a folder on the shelf of the site cabin – but something we create between ourselves on sites on a daily basis.

Practitioners are the ones who practice, and so they are the ones who are best placed to bring about any change.

However, throughout the process of unpacking construction site safety, some fundamental suggestions for change have emerged.

For example, that safety is not easy must be recognised and given due consideration. Safety is not something that can be effectively managed through a set of folders in an office, with an hour or so in the morning to get signatures on method statements and risk assessments, and to whizz through a few induction slides about the project. The way we label and isolate safety through safety management systems, safety training or safety programmes also contributes to an overly simplistic idea of safety, as something that can be easily managed as a bolt-on to the much more important business of construction. But as we have found, safety does not work like this, and to try to manage it as such is simply not working as well as it could. Instead, it is arguable that safety is something that must be embedded within work practices, including management and supervision; it is something they do, create and participate in every day. But as a consequence of this the time and efforts needed to get involved with and influence the complicated, changeable and awkward safety found on sites must be readily acknowledged and indeed provided.

We have also revealed that this is not something that can be delivered through any specific 'safety management role', separated from those whose work is actually that of operations and production. This organisational structuring simply segregates safety from the point of its existence in practice and safety doesn't work like this. To suggest more efforts be directed to this kind of management 'support' simply would not fit, and instead it is essential that safety is positioned throughout and within operations and production – but with the resources needed to do so effectively in practice.

This links closely to the practice of bundling safety up with health, the environment and quality – everything that is not directly productive grouped under one convenient heading of 'HSEQ', with the rather ambitions task of its management again given to that manager sitting outside of the practices of production. Whilst the argument can be made that this is to the detriment of *all* of these aspects of construction work (which of course brings many other consequences in terms of our impact and effectiveness as an industry), our concern here is safety – and for safety to be not only segregated but then positioned to be of equal importance to the tolerance of the brickwork or the finish of the tape jointed wall, is certainly cause for concern. Not only should these elements be un-bundled, but they should also be embedded back alongside practice – and as noted above, with the resources provided for their effective inclusion and due prioritisation alongside the more practical business of production on sites.

Another aspect revealed through this unpacking has been the *way* we talk about safety on sites. The way we label safety as safety, but also danger as safety – there is a lack of consistency that does not help to contribute to a coherent idea of safety in practice. The change inherent in construction necessitates the need for signage – but this should perhaps be given more careful thought (the resources noted above again coming into play). Consistency in information provided, the why of safety duly addressed for the workforce, and the need to mitigate 'sign fatigue' – where no sign is acknowledged as there are either too many or several out of date and now defunct – should all be carefully considered by the site team.

The way accidents are prioritised in the way we talk about safety should also be reflected upon. This focus on the events of unsafety is both distracting and unrepresentative of safety in practice, and indeed perpetuates the idea that accidents *will* happen. This has often been extrapolated through safety management programmes to the ultimate quest for zero. Zero, the biggest number in construction safety, does not resonate with the workforce, who do not believe it will ever come to pass, but does provides management with a highly distracting fixation on numbers and measurement that do not have any real relevance to safety as it is understood in practice. Zero should be repositioned, yes it should be our utopia, but it should not become a replacement for safety in practice. Zero is the stuff of corporate social responsibility propaganda, and should not really be so prominent a part of safety on sites.

Instead, perhaps we would do worse than to reconsider safety as the non-event and refocus on safety in practice – or rather consider the re-prioritisation of risk and danger over safety, as the earliest legislation around safety and industry sought to do. If we rethink safety in this way we are able to again embed it in the work and management of that work, rather than segregate it from practice and garnish it with slogans and big round logos.

Personal protective equipment is also something that should be reconsidered. It is often unduly prioritised *as* safety, but is actually something which has the least influence in how safety manifests in practice. Whether the continued layering of the site workforce in PPE is beneficial, or is simply degrading the relevance and importance of *any* PPE requires serious thought. Some PPE doesn't work, it is not helpful and therefore is resented. Whether it does bring wider benefit should be carefully assessed, to avoid it becoming another nonsense put in place in the 'interests of safety' (Brown and Hanlon 2014). This seeps into our wider understandings around safety, and enables the workforce to readily deride *all* PPE, and *all* safety rules. We should also stop talking about safety through PPE – and instead try to educate and articulate with direct and specific reference to practice and work – which will in turn help to embed ideas of safety into practice, rather than leave them all too readily associated with hard hats, glasses and gloves – which are simply the artefacts of safety we use on sites.

Although legislation forms a core mechanism for the management of safety, it has also supported a highly inflexible and polarised approach – but safe/unsafe simply does not work like that in reality. Yet how to monitor fluidity and change effectively remains a challenge, although there are of course those best placed to do so – those working to bring about that change as they place bricks, erect scaffolding and pour concrete. Yet to effectively embed safety within the practices of work that lead to these changes has also been revealed throughout this book to be problematic – and certainly more complex than the way the simplistic term 'workforce engagement', so frequently used on sites, would suggest. It is much harder to make a reality, and it is not helped by the various layered hierarchies, or safety cultures, found within the site environment. Different voices speaking very different safety languages have been identified and revealed – the no-blame culture of the corporate voice clashing with the site management need for punishment, as they struggle to enforce safety within a reality in which violations of the workforce – who readily expect to be punished when (or rather if) they are caught – are so common they are simply just another accepted way of understanding safety on sites. This incoherence is further compounded by corporate safety propaganda presenting slogans and images seeking to 'make safety personal' – which might seem like the ideal solution to workforce engagement – but is often done without the realisation that this also makes risk and danger personal. But you cannot tell a person to take responsibility for their own safety without also making them responsible for their own unsafety – and this is something that is directly influenced by what is arguably the biggest challenge to safety in construction: the construction site context.

It is the ongoing struggle between productivity and safety that arguably holds the most influence. The prioritisation of time and money, through elongated supply chains from client to sole trader, held together by fierce contractual bonds, is integral to how safety works on sites. The site workforce knows this all too well – that is why supervisors, foremen and operatives so easily and readily position safety violations as an inherent part of safety on sites. Safety violations are simply something that needs to be done to get the work completed within the parameters of time and money imposed on their work practices. They are a fundamental part of how it works, and it is delusional to think otherwise.

Whilst we like to tell people to prioritise safety within their work, if the realities of the work environment do not permit or support this, then it becomes yet more empty safety rhetoric. All the safety propaganda slogans and targets and engagement programmes in the world will not have any effective impact if safety is not *actually* what is being prioritised on site.

As many of the other practical suggestions made above have shown, alongside reflection and careful consideration of how we can try to reposition safety in practice, it is all too often about time and money – more specifically the provision and allocation of these precious resources to enable people to embed safety into their work practices. To have the time to consider change, to be able to reflect and reprioritise tasks according to the immediate work environment, and to be able to operate within the incoherence and inconsistency that is simply a part of the highly complex way safety works on sites. But this is currently something of a luxury and is unfortunately likely to remain so, given the wider operational markets and world we currently live in.

However, as was suggested about safety itself, there is also the potential for more explicit acknowledgement of this very fundamental challenge to safety on sites to be helpful in its own right. If we are able to position and accept these constraints and influences much more readily within not only the construction site environment, but also the boardrooms of industry organisations and our clients, there is then scope for change in how we understand safety on sites in its broadest possible terms. And with this comes the potential to bring about beneficial modifications to the way our industry operates at this level – we are able to move safety away from the site and the sharp end, and back to its blunt end (Hollnagel 2014) where decisions are made that affect or limit these resources. As Dekker (2006) suggests, we are able to reveal the consequences of these 'latent defects' within our wider construction system and start to think about how procurement, contracts and our use of supply chains so significantly influences safety on sites.

Whilst the way safety has been unpacked in this book has been able to offer some practical suggestions to begin to reposition safety on sites, it has also provided many insights for industry to take these ideas even further. From small changes that can be made relatively easily, to the need to perhaps rethink safety from corporate perspectives and bring about change in how safety currently works (or rather doesn't work) in terms of management structures, corporate social responsibility, and the way we procure and organise our site workforce. It is hoped that this book has proved useful, if only to make us think differently by revealing the way safety is constructed on sites.

Summary

This final chapter has sought to consolidate this unpacking of safety on sites, summarising and bringing together the many different aspects that combine to form our shared understandings of safety on sites. Safety is highly complex, and readily shifts and changes depending on people, place and time. These perspectives have led to an alternative positioning of 'safety culture' – from something measurable and almost tangible, created through policy and process, to something much better able to fit with the awkward and incoherent safety that is actually found on sites.

The insights brought to light through this unpacking of safety are not only academic, but can also readily relate to the real world – and various ideas and suggestions for interventions and changes have been drawn out, which work with the way safety itself actually works on our sites. It is also hoped that practitioners will be able to build on these ideas and new understandings to further develop safety in practice, with implementations in both industry boardrooms as well as at the site level.

As was noted in Chapter 4:

Safety matters. And it really matters to individuals and their families. But we know this.

And of course we do know this, but we may need to be more considered and nuanced in the ways we know this, and therefore the ways we go about seeking change in practice – and hopefully this book has been able to provide some support, illumination and guidance in revealing what safety is, and how it actually works on construction sites.

Of course the way we will know if we have ever been truly successful about safety in construction is when we stop talking about it at all. We won't even mention it. When it is no longer a segregated entity, something we need to prioritise and prominently label – but instead is simply what we do and how we do it when we carry out our work in the construction industry.

References

Antonsen, S. (2008) *Safety Culture – Theory, Method and Improvement.* Ashgate, Farnham.

Brown, T. and Hanlon, M. (2014) *In the Interests of Safety: The Absurd Rules That Blight Our Lives and How We Can Change Them.* Sphere, London.

Choudhry, R.M., Fang, D. and Mohamed, S. (2007) The nature of safety culture: a survey of the state-of-the-art. *Safety Science,* **45**(10), 993–1012.

Clarke, S. (2006) Contrasting perceptual, attitudinal and dispositional approaches to accident involvement in the workplace. *Safety Science,* **44**(6), 537–50.

Dekker, S. (2006) *The Field Guide to Understanding Human Error.* Ashgate, Farnham.

Denison, D.R. (1996) What is the difference between organisational culture and organisational climate? A native's point of view on a decade of paradigm wars. *Academy of Management Review,* **21**(3), 619–54.

Dov, Z. (2008) Safety climate and beyond: a multi-level multi-climate framework. *Safety Science,* **46**(3), 376–87.

Edwards, J.R.D., Davey, J. and Armstrong, K. (2013) Returning to the roots of culture: a review and re-conceptualisation of safety culture. *Safety Science,* **55**, 70–80.

Fang, D., Wu, H. (2013) Development of a Safety Culture Interaction (SCI) model for construction projects. *Safety Science,* **57**, 138–49.

Glendon, A.I. (2008) Safety culture and safety climate: how far have we come and where could we be heading? *Journal of Occupational Health and Safety Australia and New Zealand,* **24**(3), 249–71.

Guldenmund, F.W. (2007) The use of questionnaires in safety culture research – an evaluation. *Safety Science,* **45**(6), 723–43.

Hale, A. (2000) Culture's confusions. *Safety Science,* **34**(1-3), 1–14.

Hollnagel, E. (2014) *Safety I and Safety II – The Past and Future of Safety Management*. Ashgate, Farnham.

Hopkins, A. (2002) *Safety Culture, Mindfulness and Safe Behaviour: Converging Ideas?*, Working Paper 7, National Research Centre for OHS Regulations, The Australian National University.

Townsend, A.S. (2013) *Safety Can't Be Measured: An Evidence-based Approach to Improving Risk Reduction*. Gower, Farnham.

Wamuziri, S. (2011) Factors that contribute to positive and negative health and safety cultures in construction. In *Proceedings of the CIB W099 Conference Prevention – Means to the End of Construction Injuries, Illnesses and Fatalities*. CIB, Rotterdam.

Chapter Ten
Reflections

The idea that first inspired this project was very simple – I was going to find out why people still had accidents on large construction sites, despite there being lots of management efforts made to prevent this, and I was going to sort this out by working out how to develop a positive construction site safety culture.

With hindsight this was of course far too simple, optimistic and in fact pretty impossible.

What I found instead was an inconsistent, incomplete and often incoherent safety; a safety that often didn't make sense, contradicted itself and was ever changing depending on time, space and place. But actually, this worked for me. As someone who has spent a long time on site, and for whom a key part of everyday was involved with enforcing the site safety rules whilst also attempting to engage with people and convince them why safety was important, these mixed up and messy ideas of what safety was really spoke to me. They explained why things were as they were, why safety was as it was, and were able to fit really well with how my world on site worked.

And as such I was encouraged to continue my research, to ask questions and talk to people and explore and to eventually share them through this book. I hope it has been delivered in such a way that it is able to resonate with those who also have experience working on sites, but also with those who research safety more academically – and who might wonder why, despite

Unpacking Construction Site Safety, First Edition. Dr Fred Sherratt.
© 2016 John & Wiley Sons, Ltd. Published 2016 by John & Wiley Sons, Ltd.

all the questionnaires, critical factor analysis and models and management processes that have been developed over many years, safety still doesn't really work for construction.

I would also like to highlight that the safety of this book is not all bad – safety does exist as practice, people do try to prioritise it, take responsibility for it, and it is valued and held dear by many of those who work on sites – but of course not all the time, the world doesn't work like that. And more significantly, problems do arise when time and money are put into the mix, and all too often it is safety that loses out. However, with the new understandings this book brings to safety research in the construction industry, I would hope that these considerations will be given a much higher priority by clients, contractors and their supply chains, to take us closer to a more coherent and integrated understanding of what safety actually *is* on site, and enable us to do something about it that actually *works* in practice.

Index